SUPPORT VECTOR MACHINES

and **Their** **Application** in **Chemistry** and **Biotechnology**

SUPPORT VECTOR MACHINES

and **Their** **Application** in **Chemistry** and **Biotechnology**

Yizeng Liang • Qing-Song Xu
Hong-Dong Li • Dong-Sheng Cao

CRC Press
Taylor & Francis Group
Boca Raton London New York

CRC Press is an imprint of the
Taylor & Francis Group, an **informa** business

CRC Press
Taylor & Francis Group
6000 Broken Sound Parkway NW, Suite 300
Boca Raton, FL 33487-2742

First issued in paperback 2018

© 2011 by Taylor and Francis Group, LLC
CRC Press is an imprint of Taylor & Francis Group, an Informa business

No claim to original U.S. Government works

ISBN-13: 978-1-4398-2127-5 (hbk)
ISBN-13: 978-1-138-38197-1 (pbk)

Library of Congress Cataloging-in-Publication Data

Support vector machines and their application in chemistry and biotechnology / Yizeng Liang ... [et al.].
    p. cm.
Includes bibliographical references and index.
ISBN 978-1-4398-2127-5 (hardback)
1. Support vector machines. 2. Chemometrics. I. Liang, Yizeng.

Q325.5.S866 2011
542'.85--dc22
2010041498

Visit the Taylor & Francis Web site at
http://www.taylorandfrancis.com

and the CRC Press Web site at
http://www.crcpress.com

Contents

Preface

"Make everything as simple as possible, but not simpler."

Albert Einstein

Developed for pattern recognition and later extended to multivariate regression, support vector machines (SVMs) were originally proposed by Vapnik et al. and seem a very promising kernel-based machine-learning method. What distinguishes SVMs from traditional learning methods, in our opinion, lies in their exclusive objective function, which minimizes the structural risk of the model. The introduction of the kernel function into SVMs made the method extremely attractive, because it opened a new door for chemists and biologists to use SVMs to solve difficult nonlinear problems in chemistry and biotechnology through the simple linear transformation technique. The distinctive features and excellent empirical performance of SVMs have drawn chemists and biologists so much that a number of papers, mainly concerned with the applications of SVMs, have been published in chemistry and biotechnology in recent years. These applications cover a large range of meaningful chemical or biological problems, for example, spectral calibration, drug design, quantitative structure–activity and property relationships (QSAR/QSPR), food quality control, chemical reaction monitoring, metabolic fingerprint analysis, protein structure and function prediction, microarray data-based cancer classification, and so on.

However, we should also admit that SVMs are not as widely used as traditional methods such as principal component analysis (PCA) and partial least squares (PLS) in chemistry and metabolomics. In order to efficiently apply this rather new technique to solve difficult problems in chemistry and biotechnology, one should have a sound in-depth understanding of what kind of information this new mathematical tool could really provide and what its statistical properties are. However, the difference in professions makes one feel worlds apart. The gap between the mathematicians and the chemists as well as biologists is actually very large because most chemists and biologists are not quite familiar with the theoretical mathematical language and its abstract descriptions. Undoubtedly, this gap limits the applications of SVMs. It goes without saying that the deeper the understanding of SVMs one has, the better application one may achieve. However, to our best knowledge, there is currently no book that provides chemists and biologists with easy-to-understand materials on what SVMs are and how they work.

Thus it seems urgent to build a much needed bridge between the theory and applications of SVMs and hence lessen and even fill the gap. This book aims at giving a deeper and more thorough description of the mechanism of SVMs from the point of view of chemists and biologists and hence making it easy for those scientists to understand. We believe that we might have found the way to do this. Thus it could be expected that more and more researchers will have access to SVMs and further apply them in the solution of meaningful problems in chemistry and biotechnology. We would like to say the above discussion is our main motivation in writing such a book to construct a bridge between the theory and applications of SVMs.

This book is composed of eight chapters. The first four chapters mainly address the theoretical aspects of SVMs, and the latter four chapters are focused on the applications on the quantitative structure–activity relationship, near-infrared spectroscopy, traditional Chinese medicines, and OMICS studies, respectively.

We would like to acknowledge the contributions of many people to the conception and completion of this book. Lance Wobus of Taylor & Francis Books/CRC Press is gratefully acknowledged for his kind encouragement in writing this book as well as his great help which enabled its publication. Dr. Wei Fan in our group is also acknowledged for providing the near-infrared data as well as the helpful discussions on data analysis. Dr. Liang-Xiao Zhang in our group is acknowledged for his help with the literature.

Yizeng Liang
Changsha, Yuelu Mountain

Author Biographies

Yizeng Liang

Professor Yizeng Liang earned his PhD in analytical chemistry in 1988 from Hunan University, China. From June 1990 to October 1992, he worked at the University of Bergen, Norway on a postdoctoral fellowship from the Royal Norwegian Council for Scientific and Industrial Research (NTNF). In 1994, he received a DPhil from the University of Bergen, Norway. He is now a professor of analytical chemistry and chemometrics, doctoral supervisor, director of the Research Centre of Modernization of Traditional Chinese Medicines, and vice dean of the College of Chemistry and Chemical Engineering, Central South University, China. He is also a council member of the Chemical Society of China; a member of the Analytical Chemistry Commission, Chemical Society of China (since 1995); vice chairman of the Computer Chemistry Committee, Chemical Society of China (since 2001); member of the advisory editorial boards of the international journals *Chemometrics and Intelligent Laboratory Systems* (since 1998), *Near Infrared Analysis* (since 2001), *Journal of Separation Science* (since 2005), and the *Chinese Journal of Analytical Chemistry* (*Fenxi Huaxie*, since 1995); as well as editor of *Chemometrics and Intelligent Laboratory Systems* (since 2007).

Since 1989, Professor Liang has published more than 360 scientific research papers, over 300 of which were published in the source journals of the Scientific Citation Index (SCI) with an *h*-index of 30. In addition, he has published eight books (seven in Chinese and one in English) and has authored six chapters in three English-language books.

Dr. Liang has received a number of awards from various ministries of the Peoples' Republic of China and Hunan Province for his research on chemometric methods for white, gray, and black analytical systems (1994); multicomponent analytical systems (1995); complex multicomponent systems and computer expert systems for structure elucidation (2002 and 2003); and multivariate methods for complex multicomponent systems and their applications to traditional Chinese medicine (2009).

Professor Liang's research interests include analytical chemistry, chemometrics, chemical fingerprinting of traditional Chinese medicines, data mining in chemistry and Chinese medicines, metabolomics, and proteomics, among others.

Qing-Song Xu

Qing-Song Xu is a professor at the Institute of Probability and Mathematical Statistics, School of Mathematical Sciences and Computing Technology, Central South University, China. He obtained his PhD in applied mathematics from Hunan University, China, in 2001.

Qing-Song Xu's main research contributions have been in the field of applied statistics and chemometrics, and he has written dozens of papers in these areas. His current research focuses on applied problems in chemistry, biology, and medicine, in particular data analysis, classification, and prediction problems.

chapter one

Overview of support vector machines

Contents

1.1 Introduction

Machine learning is a scientific discipline related to the design and development of algorithms that allow computers to learn from the given training data. Generally, the learning algorithms can be classified into two taxonomies: unsupervised learning and supervised learning, according to whether an output vector is needed to supervise the learning process. Supervised learning can be further divided into two types: regression and classification. The former refers to the situation where the output vector consists of continuous value and the latter refers to the case where the output vector denotes the discrete class label of each sample. This book introduces the state-of-the-art supervised learning algorithm, and support vector machines (SVMs), as well as its applications, in chemistry and biotechnology. A huge amount of data, such as vibration spectra, drug activity, OMICS, data from analytical instruments, and microarray experiment-based gene expression profiles have recently been generated in chemistry and biotechnology. How to mine useful information from such data using SVMs is the main concern of this book. In this chapter, the background and key elements of SVM together with some applications in chemistry and biotechnology are briefly described.

1.2 Background

SVMs, developed by Vapnik and his coworkers in the field of computer science, are supervised machine learning algorithms for data mining and knowledge discovery. They stem from the framework of statistical learning theory or Vapnik–Chervonenkis (VC) theory and were originally developed for the pattern recognition problem. To date, VC theory is the most successful tool for accurately describing the capacity of a learned model and further telling us how to ensure the generalization performance for future samples by controlling model capacity. The theory mainly concerns the consistency of a learning process the rate of convergence of a learning process, how to control the generalization performance of a learning process, and how to construct learning algorithms. VC dimension and the structural risk minimization (SRM) principle are the two most important elements of VC theory. VC dimension is a measure of the capacity of a set of functions and the SRM principle can ensure that the learned model can generalize well.

Historically, all the necessary elements that form the theory and algorithm of SVMs have been known since the early 1970s. But it took about 25 years before the concept of SVMs was developed and the spirit of SVMs was systematically elucidated in a formal way in the two fundamental monographs: *Statistical Learning Theory* and *The Nature of Statistical Learning Theory* [1, 2]. As pointed out in the two books, in contrast to traditional learning methods where dimension reduction is performed in order to control the generalization performance of the model, the SVMs dramatically increase the dimensionality of the data and then build an optimal separating hyperplane in the high-dimensional feature space relying on the so-called margin maximizing technique. It's surprising but expected that very excellent performance is observed when SVMs are applied to practical problems such as handwriting recognition.

The distinctive features and excellent empirical performance greatly accelerate the expansion of the SVM idea and further lead to their application in a wide variety of fields, such as credit rating analysis [3], text classification [4], spectral calibration [5,6], QSAR/QSPR [7,8], drug design [9], cancer classification [10], protein structure and function prediction [11,12] and metabolomics [13]. All these contribute to both the theoretic and experimental development of SVM and make it a very active area.

Considering the outstanding performances of SVMs, one may be eager to discover what on earth makes SVMs so powerful. To this end, it's first recommended to have a basic mastering of the necessary mathematics that formulate the theory of SVM, which include, but are not limited to, maximal interval linear classifier, kernel functions, kernel matrix, feature spaces, optimization theory (linear or quadratic programming), dual representations, and so on. Now let's go through these key points step by step.

1.2.1   Maximal interval linear classifier

In classification, the simplest model is the linear classifier, which was mainly developed in the precomputer age of statistical and machine learning. Even in the current era with rapid computer development, there are still enough reasons to employ the linear model to perform our studies because it is easy for us to understand the latent input/output relationship and to make some further interpretations or statistical inferences based on the established linear model.

Linear regression is the simplest way to build a linear classifier. Typically we are given a dataset of m samples collected into a matrix X of size $m \times p$, where p denotes the number of variables. The class label vector is y with its element "1" standing for the positive class or "–1" standing for the negative class in the binary classification setting. The linear regression model has the following formula:

$$y = X\beta + e \tag{1.1}$$

where β is the unknown regression coefficient vector and e denotes the systematic error. By minimizing the squared error loss function which is in vector form defined as

$$RSS(\beta) = (y - X\beta)^T (y - X\beta) \tag{1.2}$$

where RSS stands for residual sum of squares, the least squares solution to the linear model can be easily computed as

$$\hat{\beta} = (X^T X)^{-1} X^T y \tag{1.3}$$

Then one can make predictions using the following fitted model.

$$\hat{y} = X\hat{\beta} \tag{1.4}$$

Given a new sample x_{new} which has not been seen by the fitted model shown in Equation (1.4), let's see how to predict its class. If the computed $\hat{y}_{new}$ by Equation (1.4) is positive, then one can say that x_{new} should belong to the positive class "1". On the contrary, if $\hat{y}_{new}$ is a minus number, the class of x_{new} should be predicted as "–1". The prediction rule can be summarized as

$$Predicted\ class\ label = \begin{cases} 1, & if\ \hat{y}_{new} > 0 \\ -1, & if\ \hat{y}_{new} > 0 \end{cases} \tag{1.5}$$

Let's see an example. We first simulate a two-dimensional dataset of 22 samples, with 15 samples belonging to the positive class (circle marker) and the remaining 7 samples to the negative class (diamond marker). The data are shown in Figure 1.1A. Then a linear regression model is fitted to the data and the resulting linear classifier is also given in Figure 1.1A as a solid line. Apparently in this linearly separable case, all the samples are correctly classified by the constructed linear classifier. The careful reader will find that the classifier is very close to the positive class and may further claim that this classifier is not reliable because the positive sample is easily misclassified as a negative sample if it is contaminated by noise. In other words, this classifier is unfair to the positive samples and somewhat "dangerous." Therefore, it seems necessary for us to develop a "safer" strategy based on which one can establish a "safer" or more reliable classification rule. Next we address this "safer" kind of classifier: a linear classifier with maximal interval.

Figure 1.1B shows the same data as shown in Figure 1.1A. Note that there are altogether three parallel lines. Two dashed lines are located on the boundaries of the two classes of samples and the solid line is in the middle of the two dashed lines. Further suppose that the line in the middle is a candidate classifier. With these assumptions, we can now define the interval of the candidate classifier as the distance between the two dashed lines. Intuitively, this definition has a clear geometrical explanation and is very easy to understand. Naturally, the one whose interval achieves the maximum is defined as the linear classifier with maximal interval. By the way, it should be mentioned here that the interval of a

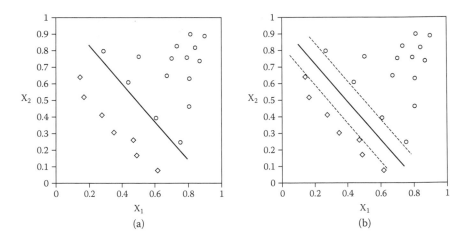

Figure 1.1 Illustration of the linear classifier built by using linear regression (Plot A) and the linear classifier with maximal interval (Plot B).

classifier can also be termed *margin*. In the SVM research, we prefer to use the term *margin* instead of interval.

Compared to the linear classifier in Figure 1.1A, one can easily find that the linear classifier with maximal interval should be the best choice in that it has the larger capacity to tolerate the noise or error. Intuitively, it is the safest one. Of course, it is necessary for us to know how the maximal interval classifier can be computed because it is very important for developing SVMs. We can perhaps call it the predecessor of SVMs. However, we do not show the computational procedure here. Readers can refer directly to Chapter 2 for detailed information on both mathematics and computations.

Let's reconsider the example in Figure 1.1. In this case, the simulated data are linearly separable and the margin (interval) between classes can be easily defined in geometrics. However, the reader may ask questions such as, "Does the notion of maximal interval linear classifier still work if the data are linearly inseparable? If it does, where is the linear classifier and what is the margin?" It is not possible for us to find such a linear classifier in the exterior, let alone the one with maximal interval. However, it is shown by taking a simple but enlightening example where the linear classifier does exist in the so-called feature space produced by kernel functions: another key element for developing SVM.

1.2.2   *Kernel functions and kernel matrix*

To intuitively understand the notion of a kernel function, let's first see another dataset shown in Figure 1.2. The data in each class are distributed in two circular regions. Each sample belongs to either the positive class (plus marker) or negative class (asterisk marker) and they cannot be separated well using a linear classifier in the 2-D space (Figure 1.2A). One way to solve this problem is to construct very complicated nonlinear models, for example, ANN. However, it should be noticed that the adjustment of tuning parameters of such kinds of model is usually a time-consuming and tedious task. Moreover, the learned nonlinear discriminating function is most often so flexible that it is difficult for one to ensure its generalization performance. However, the other feasible and effective solution is simply to increase the dimension of the data.

In this case, let's increase the dimension of each sample by one. For the ith sample $\mathbf{x}_i = [x_{i1}, x_{i2}]$, the value of the third dimension can simply be calculated as $x_{i3} = x_{i1}^2 + x_{i2}^2$. Thus, in the 3-D space in Figure 1.2B, the ith sample can be denoted by $\mathbf{x}_i = [x_{i1}, x_{i2}, x_{i3}]$. Indeed, this operation realizes a nonlinear mapping of the original data from the input space (original low-dimensional variable space before increasing dimensions) into feature space (higher-dimensional space after increasing dimensions). This operation, usually called feature mapping, is primarily an implicit

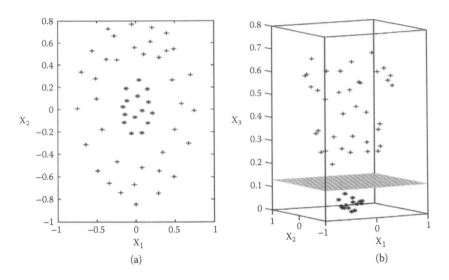

(a)   (b)

Figure 1.2 The two plots ((A) and (B)) show that the two classes of linearly inseparable samples (plus and asterisk) in a two-dimensional space can be separated linearly without errors when the third dimension is added to each sample. It should be mentioned that the operation of increasing data dimension can be easily and explicitly implemented by using the kernel function.

characteristic of the kernel function. Obviously, it can be seen that the linearly inseparable samples in 2-D space can be separated by a linear hyperplane without any errors.

But we face some problems. For instance: how could we efficiently choose the function to compute the additional dimension? Can we ensure the dimension-increased data are linearly separable? Do we have to know explicitly the function for adding dimensions? We illustrate that the kernel function does provide a smart solution to this problem. It serves as a dimension-increasing technique and further transforms the linearly inseparable data into linearly separable data in the feature space. More interesting, with the help of kernel functions, it's not even necessary for us to know the mathematical form of the functions for adding dimension. However, kernel function details are not delineated in this chapter. But a brief introduction is necessary. Please consult Chapters 2 and 3 for both theoretical and computational details for kernel functions.

Briefly, a kernel function is primarily a symmetric mathematical function that has the general form shown in Equation (1.6),

$$K(X_i, X_j) = <\varphi(X_i), \varphi(X_j)>, i, j = 1, 2, 3 ... m \tag{1.6}$$

where $<\cdot>$ denotes the inner product and $\varphi(\cdot)$ is a set of mapping functions that can project the original samples into a high-dimensional feature space. That is to say, $\varphi(X_i)$ is a vector of higher dimension than X_i. Furthermore, for each pair of samples, one can compute an inner product using Equation (1.6). All the inner products are now collected into a matrix $\mathbf{K}$ with elements

$$\mathbf{K}_{ij} = K(X_i, X_j) \tag{1.7}$$

This matrix $\mathbf{K}$ is the so-called kernel matrix, which without any exception is a key point of all the kernel-based algorithms. From this perspective, SVMs are just a special case of kernel methods. As is known, the inner product is a measurement of the similarity between two samples. In this sense, each element of the kernel matrix reflects the similarity between the samples in the feature space produced by $\varphi(\cdot)$.

So far, we have at least a basic understanding of kernel functions and the associated kernel matrix. Now it is time for us to determine what properties of a function $K(X_i, X_j)$ are necessary to ensure that it is a kernel function for some feature space. According to Mercer's theorem, assume that X is a finite input space with a symmetric function on X. It becomes a kernel function if and only if the resulting kernel matrix $\mathbf{K}$ is positive semidefinitive, that is, without negative eigenvalues.

1.2.3 Optimization theory

Optimization theory is very important for SVMs because the computation of the SVM model can be converted to find the solution of a corresponding optimization problem. In the area of optimization, the most frequent cases we are confronted with may be those of constrained optimization. A special constrained case is convex optimization where the feasible solution of the optimization problem is convex, meaning that any connecting line between two points of the feasible region still falls in the region. There are many freely available programs written in C++, MATLAB®, or R for solving convex problem. But how are SVMs related to the optimization problem?

SVMs work mainly in three steps: (1) using a given kernel function to transform the original data into the feature space; (2) mathematically defining a general linearly separating hyperplane associated with a margin in the feature space, and further establishing a convex optimization problem with maximizing the margin as the objective function; and (3) finding the solution to the convex problem. The solution is just the linear classifier in the feature space with maximal margin, which is the so-called SVMs classifier. It is also called the optimal separating hyperplane (OSH).

As we know, the Lagrange multiplier method is a famous technique for solving optimization problems. In the case of SVMs, the problem of maximizing the margin can also be solved by this method. Without loss of generality, we assume that the derived optimization problem with Lagrange multipliers α introduced is

$$\text{Maximize: } f(w,\alpha) \tag{1.8}$$

Here w denotes the primal variables of a SVM model. By deriving (1.8) against w and setting it to zero, one can derive an equivalent optimization problem with w eliminated. That is,

$$\text{Maximize: } g(\alpha) \tag{1.9}$$

In (1.8) and (1.9), α is called the dual variable. Formula (1.8) is called the primal problem, and Formula (1.9) is called the dual problem corresponding to (1.8). This characteristic is the so-called duality. Only the general notion is given here. In Chapter 2, we show this in great detail.

The Lagrangian treatment of convex optimization problems results in an alternative dual representation, which in most cases turns out to be much cheaper to compute than the primal problem. The reason for this is that the dual problem is just a convex problem with simpler constraints and can be solved easily using quadratic programming (QP). By the way, dual strategies play a central role in kernel methods, such as kernelized Fisher discriminant analysis (KFDA) [14–17] and kernelized partial least squares (KPLS) [18,19], and so on. Support vector machines are special kernel methods that possess some distinguished properties, such as margin maximization and sparsity, which are discussed later.

1.3   Elements of support vector machines

In the last section, we introduced the foundations of support vector machines. Here, we string them together, trying to give an overall picture of SVMs. Figure 1.3 shows the basic procedures for computing a SVM model. Summing up, a SVM model is the mathematical solution of a convex optimization problem whose objective is to maximize the margin of the linear classifier in the feature space produced by the user-chosen kernel function. It should be emphasized here, according to William S. Noble [20], that the basic idea behind the SVM classifier can be explained without ever reading an equation. To understand the essence of the SVM classifier, one only needs to grasp four concepts: kernel function, feature space, separating hyperplane, and optimization problem. Once again, it

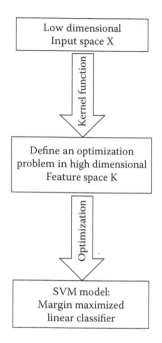

Figure 1.3 The procedure for computing an SVM model.

should be addressed here that SVMs, originally developed for classification, were later expanded and can also be used for regression and density estimation.

In addition, we feel it necessary to give a concise description of the variants of support vector machines and some related methods for the sake of a comprehensive understanding of this field. The first introduction of the SVM idea in the 1990s attracted researchers in a wide variety of areas. A number of variants were developed, most of which dwelled on how to improve the robustness of SVM models [21–23]. Later, one significant development was the establishment of least squares-support vector machines (LS-SVMs) [24–27], which are a reformulation of standard SVMs and lead to solving linear KKT (Karush–Kuhn–Tucker) systems. In the theory of optimization, the Karush–Kuhn–Tucker conditions are necessary for obtaining an optimal solution of a nonlinear programming problem. The solution of LS-SVMs is usually not sparse. Another important version is the relevance vector machine (RVM), which uses Bayesian inference to obtain sparse solutions for regression and classification. The only thing we emphasize here is that RVMs have an identical functional form to support vector machines, but provide probabilistic classification [28–30].

1.4 Applications of support vector machines

One concern of readers is how SVMs perform when dealing with real-world problems. Here we give some selected applications in the fields of chemistry and biotechnology, aiming at illustrating that SVMs are indeed a good alternative in modeling the chemical or biological data produced by modern analytical instruments.

The first application comes from the field of multivariate calibration of near-infrared (NIR) spectral data. NIR spectroscopy has been gaining broad applications in a variety of fields in that it's fast and nondestructive to samples under investigation. The most commonly used method for dealing with NIR data may be partial least squares (PLS) regression. However, one should bear in mind that nonlinear methods may be a better choice to cope with the nonlinearity existing in the real-world data. L. M. C. Buydens et al. employed LS-SVM to perform such a study in which the near-infrared spectra are affected by temperature-induced spectral variation [31]. As was described in this work, the used data were measured by ternary mixtures of ethanol, water, and 2-propanol. For these data, 19 different combinations of mole fractions were analyzed in a wavelength range of 850–1049 nm with a 1-nm resolution. Each mixture was measured at 30, 40, 50, 60, and 70°C (±0.2°C). Therefore, it could be observed that the measured spectra at different temperatures had some nonlinear variations. In their study, compared to the two-dimensional penalized regression and PLS regression, better results were obtained by SVMs. The authors concluded that SVMs can be seen as a promising technique to solve ill-posed nonlinear problems.

The development of the quantitative structure property/activity relationship (QSPR/QSAR) [32,33] model is chosen here as the second example because QSPR/QSAR has for a long time been an active area in chemistry for understanding the physical, chemical, or biological property at the molecular level, and has been extensively employed in drug design and related fields. To the best of our knowledge, the relationship between the molecular structure and the studied property/activity is very complex and unlikely to be linear. Thus it becomes necessary to approximate the latent model by resorting to nonlinear methods, of which SVMs may be an appropriate one. R. Burbidge et al. [9] conducted a QSAR study of predicting the inhibition of dihydrofolate reductase by pyrimidines using multilayer perceptrons, a radial basis function (RBF) network, "prune" network, "dynamic" network, and support vector machines. The biological activity under investigation was log (1/K), where K is the equilibrium constant for the association of the drug with dihydrofolate reductase. It was found that support vector machines outperformed all the other calibration methods used in the study in terms of five fold cross-validated errors [34,35]. Their study suggests that SVMs are data mining tools with great potential for QSAR.

Let's see another example in the "OMICS" study. In the field of genomics, the developed microarray experiment that monitors expression levels of thousands of genes has been gaining ever-increasing applications for cancer classification. The goal of cancer classification is involved in many aspects, such as predicting prognosis, proposing therapy according to the clinical situation, advancing therapeutic studies, and so on. In this context, it is very crucial to establish rules for predicting the phenotype of the tissue before any treatment is administered to the patient for the sake of avoiding unnecessary treatment. In Guyon et al. [36], support vector machines were used to perform gene selection and cancer classification. Their results are described briefly here. For the leukemia data, they yielded zero leave-one-out test error based on the two discovered genes. For the colon cancer data, using only four genes in the model resulted in a misclassification rate of 0.02. It was demonstrated in the study that SVM is quite powerful for the analysis of broad patterns of gene expression profiles.

As is known to us, traditional Chinese medicines (TCMs) and their preparations have been widely used for thousands of years in many Asian countries, such as China, Korea, Japan, and so on [37]. However, TCM has not been officially recognized in most countries, especially in western countries despite its existence and continued use over many centuries. The main reason may be that it is very difficult to conduct quality control of Oriental herbal drugs. Liang, Xie, and Chan [37] advocate the use of chromatographic fingerprints for comprehensively controlling the quality of TCM. Recently, Zhuo et al. [38] used support vector machines to classify the existing diagram data of the Chinese herbal medicine fingerprint. They also compared the performance of SVMs on several types of data to other classification methods. Their results showed that SVMs are a good choice for the classification of TCM fingerprints.

Since the introduction of the idea of SVMs, there has been an ever-growing literature on both theoretic and experimental aspects of SVMs. With this background, the review of SVMs becomes necessary, especially for SVM beginners or practitioners. C. Burges reviewed the support vector classification machines [39] and A. J. Smola also gave a review of the support vector regression machines [40]. Specifically, because SVMs are gaining more and more popularity in a wide variety of biological applications, W. S. Noble described the key points of SVMs and showed how SVMs work mainly for biologists [20]. Besides, Xu, Zomer, and Brereton also dwelled on SVMs for classification in chemometrics [41]. Recently, Li, Liang, and Xu in their tutorial [5] gave a comprehensive introduction of support vector machines on both classification and regression and addressed the optimization of tuning parameters of SVM using a genetic algorithm (GA).

References

1. Vapnik, V. 1998. *Statistical Learning Theory.* New York: Wiley.
2. Vapnik, V. 1999. *The Nature of Statistical Learning Theory,* second edition, New York: Springer.
3. Huang, Z., Chen, H., Hsu, C.-J., Chen, W.-H., and Wu, S. 2004. Credit rating analysis with support vector machines and neural networks: A market comparative study. *Decis. Support Syst.*, 37(4):543–558.
4. Simon, T. and Daphne, K. 2002. Support vector machine active learning with applications to text classification. *J. Mach. Learn. Res.*, 2:45–66.
5. Li, H.-D., Liang, Y.-Z., and Xu, Q.-S. 2009. Support vector machines and its applications in chemistry. *Chemometr. Intell. Lab.*, 95:188–198.
6. Thissen, U., Pepers, M., Ustun, B., Melssen, W.J., and Buydens, L.M.C. 2004. Comparing support vector machines to PLS for spectral regression applications. *Chemometr. Intell. Lab.*, 73(2):169–179.
7. Liu, H., Yao, X., Zhang, R., Liu, M., Hu, Z., and Fan, B. 2006. The accurate QSPR models to predict the bioconcentration factors of nonionic organic compounds based on the heuristic method and support vector machine. *Chemosphere*, 63(5):722–733.
8. Fatemi, M.H. and Gharaghani, S. 2007. A novel QSAR model for prediction of apoptosis-inducing activity of 4-aryl-4-H-chromenes based on support vector machine. *Bioorgan. Med. Chem.*, 15(24):7746–7754.
9. Burbidge, R., Trotter, M., Buxton, B., and Holden, S. 2001. Drug design by machine learning: Support vector machines for pharmaceutical data analysis. *Comput. Chem.*, 26(1):5–14.
10. Liu, Y. 2004. Active learning with support vector machine applied to gene expression data for cancer classification. *J. Chem. Inf. Comp. Sci.*, 44(6):1936–1941.
11. Chen, C., Zhou, X.B., Tian, Y.X., Zou, X.Y., and Cai, P.X. 2006. Predicting protein structural class with pseudo-amino acid composition and support vector machine fusion network. *Anal. Biochem.*, 357(1):116–121.
12. Cai, C.Z., Wang, W.L., Sun, L.Z., and Chen, Y.Z. 2003. Protein function classification via support vector machine approach. *Math. Biosci.*, 185(2):111–122.
13. Mahadevan, S., Shah, S.L., Marrie, T.J., and Slupsky, C.M. 2008. Analysis of metabolomic data using support vector machines. *Anal. Chem.*, 80(19):7562–7570.
14. Xu, Y., Yang, JyJ.-y., and Yang, J. 2004. A reformative kernel Fisher discriminant analysis. *Pattern Recogn.*, 37(6):1299–1302.
15. Pan, J.-S., Li, J.-B., and Lu, Z.-M. 2008. Adaptive quasiconformal kernel discriminant analysis. *Neurocomputing*, 71(13–15):2754–2760.
16. Louw, N. and Steel, S.J. 2006. Variable selection in kernel Fisher discriminant analysis by means of recursive feature elimination. *Comput. Statist. Data Anal.*, 51(3):2043–2055.
17. Li, J. and Cui, P. 2009. Improved kernel Fisher discriminant analysis for fault diagnosis. *Expert Syst. Appl.*, 36(2, Part 1):1423–1432.

18. Walczak, B. and Massart, D.L. 1996. Application of radial basis functions—Partial least squares to non-linear pattern recognition problems: Diagnosis of process faults. *Anal. Chim. Acta,* 331(3):187–193.
19. Walczak, B. and Massart, D.L. 1996. The radial basis functions—Partial least squares approach as a flexible non-linear regression technique. *Anal. Chim. Acta,* 331(3):177–185.
20. Noble, W.S. 2006. What is a support vector machine? *Nat. Biotechnol.,* 24:1565–1567.
21. Suykens, J.A.K., De Brabanter, J., Lukas, L., and Vandewalle, J. 2002. Weighted least squares support vector machines: Robustness and sparse approximation. *Neurocomputing,* 48:85–105.
22. Trafalis, T.B. and Gilbert, R.C. 2006. Robust classification and regression using support vector machines. *Eur. J. Oper. Res.,* 173(3):893–909.
23. Hu, W.J. and Song, Q. 2004. An accelerated decomposition algorithm for robust support vector machines. *IEEE Trans. Circuits Syst. II-Express Briefs,* 51(5):234–240.
24. Suykens, J.A.K. and Vandewalle, J. 1999. Least squares support vector machine classifiers. *Neural Process. Lett.,* 9(3):293–300.
25. Arun Kumar, M. and Gopal, M. 2009. Least squares twin support vector machines for pattern classification. *Expert Syst. Appl.,* 36(4):7535–7543.
26. Borin, A., Ferrao, M.F., Mello, C., Maretto, D.A., and Poppi, R.J. 2006. Least-squares support vector machines and near infrared spectroscopy for quantification of common adulterants in powdered milk. *Anal. Chim. Acta,* 579(1):25–32.
27. Chuang, C.C., Jeng, J.T., and Chan, M.L. 2008. Robust least squares-support vector machines for regression with outliers. *2008 IEEE International Conference on Fuzzy Systems, Vols. 1–5* 2008:312–317.
28. Tipping, M.E. 2001. Sparse Bayesian learning and the relevance vector machine. *J. Mach. Learn. Res.,* 1(3):211–244.
29. Majumder, S.K., Ghosh, N., and Gupta, P.K. 2005. Relevance vector machine for optical diagnosis of cancer. *Lasers Surg. Med.,* 36(4):323–333.
30. Wei, L.Y., Yang, Y.Y., Nishikawa, R.M,, Wernick, M.N., and Edwards, A. 2005. Relevance vector machine for automatic detection of clustered microcalcifications. *IEEE Trans. Med. Imag.,* 24(10):1278–1285.
31. Thissen, U., Ustun, B., Melssen, W.J., and Buydens, L.M.C. 2004. Multivariate calibration with least-squares support vector machines. *Anal. Chem.,* 76(11):3099–3105.
32. Cronin, M.T.D. and Schultz, T.W. 2003. Pitfalls in QSAR. *J. Mol. Struc. Theochem.,* 622(1–2):39–51.
33. Guha, R. and Jurs, P.C. 2004. Development of QSAR models to predict and interpret the biological activity of artemisinin analogues. *J. Chem. Inf. Comp. Sci.,* 44:1440–1449.
34. Wold, S. 1978. Cross-validatory estimation of the number of components in factor and principal component analysis. *Technometrics,* 20(397–405).
35. Stone, M. 1974. Cross-validatory choice and assessment of statistical predictions. *J. Roy. Statist. Soc. B,* 36:111–147.
36. Guyon, I., Weston, J., Barnhill, S., and Vapnik, V. 2002. Gene selection for cancer classification using support vector machines. *Mach. Learn.,* 46(1):389–422.

37. Liang, Y.-Z., Xie, P., and Chan, K. 2004. Quality control of herbal medicines. *J. Chromatogr. B*, 812(1–2):53–70.
38. Zhuo, Z.Q., Du, J.W., Liu, Q.J., and Ma, C.Y. 2008. Application of the support vector machine algorithm in distinguishing Chinese herbal medicine fingerprint diagram data. In: *Proceedings of 2008 International Conference on Wavelet Analysis and Pattern Recognition, Vols. 1 and 2.* pp. 474–479.
39. Burges, C. 1998. A tutorial on support vector machines for pattern recognition. *Data Min. Knowl. Disc.*, 2:121–167.
40. Smola, A.J. and Scholkopf, B. 2004. A tutorial on support vector regression. *Statist. Comput.*, 14:199–222.
41. Xu, Y., Zomer, S., and Brereton, R. 2006. Support vector machines: A recent method for classification in chemometrics. *Crit. Rev. Anal. Chem.*, 36:177–188.

chapter two

Support vector machines for classification and regression

Contents

2.1 Introduction

As mentioned in Chapter 1, support vector machines (SVMs) are a promising supervised machine learning method originally developed for pattern recognition problems based on the optimization problem that minimizes the structural risk of a classification model. It is also known that supervised learning can be classified into two types: classification and regression based on whether the value of the output vector is discrete or continuous. In this sense, SVMs can be divided into two categories: support vector classification (SVC) machines and support vector regression (SVR) machines. According to this classification, the basic elements and algorithms of SVC and SVR are first discussed both theoretically and experimentally in detail, respectively. Then two simulated datasets are employed to investigate the predictive performance of SVM. It is illustrated that SVMs can deal well with data of some nonlinearity. The applications of SVM to real-world data can be found in Chapters 5 through 8, respectively.

2.2 Kernel functions and dimension superiority

Generally, real-world data are usually so complex that they cannot be well modeled by means of linear machines. This means that the computational power of linear learning machines is limited. Historically speaking, the limitation of linear machines was addressed in the 1960s by Minsky and Papert [1]. To increase the computational power, multiple layers of thresholded linear functions were proposed as a solution, which led to the development of multilayer neural networks and the corresponding learning algorithms, for example, the error back propagation strategy for training this type of model [1].

The kernel function offers an alternative solution by projecting the data into a high-dimensional feature space to increase computational power. The linearly inseparable data in the original low-dimensional space could become linearly separable after being projected into the high-dimensional feature space, called dimensional superiority in our previous work [2] and also in this book. What interests us more is that the kernel function can perform nonlinear mapping to the high-dimensional feature space in an implicit manner without increasing the computational cost. Let's see now how kernel function works and how it helps to build support vector machines.

2.2.1 Notion of kernel functions

To intuitively understand the notion of a kernel function, let's consider again the simulated dataset in Figure 1.2 of Chapter 1. In this case, in order to linearly separate the two classes of data, we have first projected all the original samples into the 3-D feature space by adding an additional dimension. But how should we project the data into a space that usually has much higher dimension?

The kernel function provides a smart solution to this problem. It serves as a dimension-increasing technique and further transforms linearly inseparable data into linearly separable data. Moreover, the kernel function is simultaneously a mathematical trick to calculate the inner product in the feature space, which is a substantial step during the training of SVMs. To mathematically understand this point, let's see another example. Suppose that the original data matrix X is of order $m\times2$ (m samples, 2 variables), each sample is denoted by $x_i= [x_{i1}, x_{i2}]$. Let $\varphi(x)$ be a function set with the form:

$$\varphi(\mathbf{x}) = \{x_1^2, x_2^2, \sqrt{2}x_1x_2\} \tag{2.1}$$

which can nonlinearly map the sample in the 2-D space into the 3-D feature space. For the ith and jth sample, we have:

$$\mathbf{x_i} = [x_{i1}, x_{i2}], \ \varphi(\mathbf{x_i}) = [x_{i1}^2, x_{i2}^2, \sqrt{2}x_{i1}x_{i2}]$$
$$\mathbf{x_j} = [x_{j1}, x_{j2}], \ \varphi(\mathbf{x_j}) = [x_{j1}^2, x_{j2}^2, \sqrt{2}x_{j1}x_{j2}] \tag{2.2}$$

Their inner product is defined as

$$D(\mathbf{x_i}, \mathbf{x_j}) = x_{i1}x_{j1} + x_{i2}x_{j2} \tag{2.3}$$

And the inner product in the feature space is

$$K(\varphi(\mathbf{x}_i), \varphi(\mathbf{x}_j)) = x_{i1}^2 x_{j1}^2 + x_{i2}^2 x_{j2}^2 + \sqrt{2}x_{i1}x_{i2}\sqrt{2}x_{j1}x_{j2} = (x_{i1}x_{j1} + x_{i2}x_{j2})^2 \tag{2.4}$$

Inserting Equation (2.3) into Equation (2.4), we have

$$K(\varphi(\mathbf{x}_i), \varphi(\mathbf{x}_j)) = (D(\mathbf{x}_i, \mathbf{x}_j))^2 = (\mathbf{x}_i^t\mathbf{x}_j)^2 \tag{2.5}$$

It's surprising that the inner product, which is a measure of similarity, in the high-dimensional feature space can be calculated directly in the original input space by introducing the kernel function. In Equation (2.5), it's a polynomial kernel and has the general form:

$$K(\mathbf{x}_i, \mathbf{x}_j) = (a\mathbf{x}_i^t \mathbf{x}_j + b)^n \tag{2.6}$$

where a and b are constants and n denotes the degree of the polynomial. It has been proved that any function satisfying the Mercer conditions can be used as a kernel.

The other commonly used kernels include:

Linear kernel: $K(x_i, x_j) = ax_i^t x_j + b$
Radial basis function (RBF): $K(x_i, x_j) = \exp(-\gamma ||x_i - x_j||2)$
Sigmoid kernel: $K(x_i, x_j) = \tanh(ax_i^t x_j + b)$

For linear and sigmoid kernels, a and b are constants. For the RBF kernel, γ is a tuning parameter controlling the width of the kernel function.

Note that the kernel function provides an efficient way for the calculation of the inner product in the feature space. More interesting, with the help of kernel functions, it's not even necessary for us to know the mathematical form of the nonlinear mapping function set $\varphi(x)$ when we want to project the data into a feature space. That is, choosing a kernel function means choosing a corresponding mapping function set that is implicitly associated with that kernel. Generally, to understand the essence of kernels, one needs to grasp at least two basic points: (i) it projects the original data into a space of a much higher dimension by adding additional dimensions, and (ii) it simultaneously provides an efficient way to calculate the inner product in the feature space.

2.2.2   Kernel matrix

For a given data matrix X of size m × p, it can easily be seen that a symmetric matrix of size m × m can be obtained by computing the similarity of each pair of samples in the feature space using a given kernel function. The resulting symmetric matrix, denoted **K**, is called the kernel matrix, which is the fundamental element of all the kernel-based methods, such as radial basis function neural networks. Obviously, the kernel-transformed data **K** are not represented individually, but only through a set of pairwise inner products. Moreover, this representation does not depend on the nature of the objects to be analyzed. Any objects such as images, molecules, protein sequences, and the like can be represented in this way. Note that the kernel technique is not unique to SVM and can also be combined with other methods to form the corresponding kernel method. For example, the kernel technique can be incorporated into PLS to generate the kernel PLS (KPLS) [3,4] which can describe the nonlinearity of the data to some extent. In addition, the Fisher discriminant analysis was also kernelized for more flexible modeling [5,6].

2.3   Support vector machines for classification

Vapnik originally developed SVMs for pattern recognition. This promising method aims at minimizing the structural risk under the frame of VC theory. The SVC model is the solution to minimization of the so-called

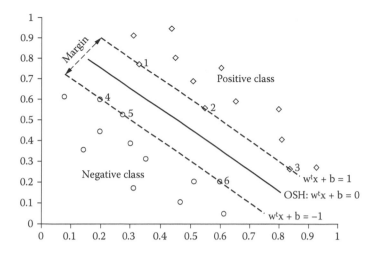

Figure 2.1 The geometrical expression of OSH and the margin for the linearly separable case. The points of diamond and circle denote the positive and negative class, respectively.

support vector machine loss function, which can be referred to in References [7–9]. Although SVCMs were initially investigated for binary classification, they have been extended to deal with the multiclass classification problem using the one-versus-all strategy [10–12]. Because the nature of multiclass classification is the same as binary classification, only the algorithm for binary classification is discussed in this section. In essence, the basic ideas of SVC can be concluded as the following in brief: (i) nonlinearly map the original data into a much higher-dimensional feature space with the help of a kernel function, and (ii) construct an optimal separating hyperplane (OSH) (Figure 2.1) in the feature space that maximizes the margin between the two classes. It's evident that such a special hyperplane is unique when data are given. The commonly seen binary classification problem can be divided into two cases, say linearly separable and linearly inseparable. The solution to the former is easy to obtain, but kernel functions have to be introduced to solve the problems in the latter case.

Suppose there is a dataset of two classes of samples, in which each sample is denoted by x_i with the corresponding class label y_i; that is,

$$x_i \in R^n, \qquad y_i \in \{-1,1\}, \qquad i = 1,2,3,\ldots,N.$$

Here, x_i is an n-dimensional vector with corresponding y_i equal to 1 if it belongs to a positive class or –1 if negative. How to derive the decision functions for the two cases mentioned above is elucidated in detail in the following section.

2.3.1   Computing SVMs for linearly separable case

For the case shown in Figure 2.1, it's apparent that the line which can classify the two classes of samples correctly is not unique. Naturally, this poses the question of which line is best. By intuition one would like to select the line in the middle because it's the "safest." This is just the most significant distinction that sets SVC apart from other classification methods. A special concept "margin," which is defined as the distance from the separating hyperplane to its nearest sample, is first introduced into the SVC model. Then, the SVC selects the hyperplane that maximizes the margin. Such a hyperplane can ensure that the SVC has good predictive ability for future samples.

In the linearly separable case, any hyperplane $f(\mathbf{x})$ should meet the condition:

$$f(\mathbf{x}_i) = \mathbf{w}^t\mathbf{x}_i + b \geq 1, \ y_i = 1 \tag{2.7}$$

$$f(\mathbf{x}_i) = \mathbf{w}^t\mathbf{x}_i + b \leq -1, \ y_i = -1 \tag{2.8}$$

where $\mathbf{w}$ is the normalized weight vector with the same dimension as $\mathbf{x}_i$ and b is the normalized bias of the hyperplane. The above two constraints can be combined into one form:

$$(\mathbf{w}^t\mathbf{x}_i + b)y_i \geq 1 \tag{2.9}$$

It should be noticed that the normalization of $\mathbf{w}$ and b makes $f(\mathbf{x})$ equal to 1 or −1 if $\mathbf{x}$ is on the boundary. Then, the margin between the two paralleled hyperplanes (Figure 2.1, broken line) can be written:

$$margin = 2\frac{|\mathbf{w}^t\mathbf{x} + b|}{||\mathbf{w}||} = 2\frac{1}{||\mathbf{w}||} = \frac{2}{||\mathbf{w}||} \tag{2.10}$$

The object of support vector classification machines is to locate the OSH that can maximize the margin subject to the constraints in inequality (2.9). Therefore, the construction of OSH can be converted to the following optimizing problem:

$$\text{maximize}: \frac{2}{||\mathbf{w}||} \tag{2.11}$$

$$\text{subject to}: (\mathbf{w}^t\mathbf{x}_i + b)y_i \geq 1$$

With the help of the Lagrange multiplier method, this problem can be further converted to minimize the objective function:

$$L(\mathbf{w}, b, \boldsymbol{\alpha}) = \frac{1}{2}\mathbf{w}^t\mathbf{w} - \sum_{i=1}^{N} \alpha_i[y_i(\mathbf{w}^t\mathbf{x}_i + b) - 1] \qquad (2.12)$$

where $\alpha_i(\alpha_i \geq 0)$ is called the Lagrange multiplier. Deriving it against $\mathbf{w}$ and b, we can obtain the two equations:

$$\frac{\partial L(\mathbf{w}, b, \boldsymbol{\alpha})}{\partial \mathbf{w}} = \mathbf{w} - \sum_{i=1}^{N} y_i\alpha_i\mathbf{x}_i = 0 \qquad (2.13)$$

$$\frac{\partial L(\mathbf{w}, b, \boldsymbol{\alpha})}{\partial b} = \sum_{i=1}^{N} y_i\alpha_i = 0 \qquad (2.14)$$

The solution to the above two equations can be easily found:

$$\mathbf{w} = \sum_{i=1}^{N} y_i\alpha_i\mathbf{x}_i \qquad (2.15)$$

$$0 = \sum_{i=1}^{N} y_i\alpha_i \qquad (2.16)$$

Inserting Equations (2.15) and (2.16) into the Lagrange function (2.12) gives us the dual form of (2.12):

$$L(\mathbf{w}, b, \boldsymbol{\alpha}) = \sum_{i=1}^{N} \alpha_i - \frac{1}{2}\sum_{i,j=1}^{N} y_iy_j\alpha_i\alpha_j\mathbf{x}_i^t\mathbf{x}_j \qquad (2.17)$$

Minimizing function (2.17) is a convex quadratic programming (QP) problem under the constraints $0 = \sum_{i=1}^{N} y_i\alpha_i$ and $\alpha_i > 0$. The optimal α_i can be calculated easily with the QP algorithm. Note that the α_i corresponding to the few samples on the boundary (Figure 2.1, broken line) are all positive and the other α_i are all equal to 0. The sample with $\alpha_i > 0$ is called the support vector (SV). Usually the support vectors occupy only a small percentage of the training data, which is called the sparsity of the solution for the minimizing function (2.17). Hence the sparsity makes the SVC model depend only on the few support vectors. The OSH is determined only by a few SVs. In Figure 2.1, there are only 6 SVs. With the computed α_i, the weight vector **w** of OSH can be obtained with Equation (2.15), and the bias can also be obtained as follows:

$$b = -\frac{1}{2}[\max_{y_i=-1}(\mathbf{w}^t\mathbf{x}_i) + \min_{y_i=1}(\mathbf{w}^t\mathbf{x}_i)] \qquad (2.18)$$

Finally, the optimized decision function (OSH) can be written in the form:

$$f(\mathbf{x}) = sgn(\mathbf{w}^t\mathbf{x}+b) = sgn\left[\left(\sum_{i=1}^{N} y_i\alpha_i\mathbf{x}_i\right)^t \mathbf{x}+b\right] \qquad (2.19)$$

2.3.2   Computing SVMs for linearly inseparable case

In most cases, the data are linearly inseparable due to noise contamination, unknown background, or intrinsic nonlinearity. As shown in Figure 2.2, it's not possible for us to construct a hyperplane that can linearly separate the two classes without error. Two techniques are introduced here to deal with this problem.

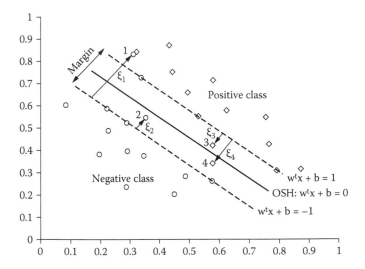

Figure 2.2 The slack variables and the OSH for the linearly inseparable case. In a similar way as the linearly separable case, the OSH can be calculated with the quadratic programming algorithm.

2.3.2.1   Slack variable-based "soft margin" technique

The slack variable was introduced by Cortes and Vapnik to construct the OSH considering the inevitable measured errors in data. The constraint inequality of the OSH can be expressed in the form:

$$(\mathbf{w}^t\mathbf{x}_i + b)y_i \geq 1 - \xi_i, \ \xi_i \geq 0, \ i = 1, 2, \ldots, N \tag{2.20}$$

where ξ_i is the slack variable which is a measure of how far the sample goes away from the boundary hyperplane (Figure 2.2). Then the construction of OSH can be expressed as the optimizing problem:

$$\text{minimize} : \frac{1}{2}||\mathbf{w}|| + C\sum_{i}^{N} \xi_i \tag{2.21}$$

$$\text{subject to} : (\mathbf{w}^t\mathbf{x}_i + b)y_i \geq 1 - \xi_i, \ \xi_i \geq 0$$

where C is a penalizing factor that controls the trade-off between the training error and the margin. With the aid of the Lagrange multiplier method and the QP algorithm (see Appendix A for more detail), the optimized solution of $\mathbf{w}$ and b can be calculated. Thus, the decision function corresponding to OSH can be written as

$$f(\mathbf{x}) = sgn(\mathbf{w}^t\mathbf{x} + b) = sgn\left[\left(\sum_{i=1}^{N} y_i\alpha_i\mathbf{x}_i\right)^t \mathbf{x} + b\right] \tag{2.22}$$

where α_i is the Lagrange multiplier for each sample.

2.3.2.2   Kernel function-based nonlinear mapping

As mentioned before, the dimensional superiority can help to solve the classification problem effectively. Naturally, the other technique to deal with the linearly inseparable case is to first nonlinearly map the original data into a much higher-dimensional feature space, then construct an OSH that maximizes the margin as in the linearly separable case. As pointed out, the kernel function can implicitly not only nonlinearly map the original data into a high-dimensional feature space but also provide a high-efficiency mathematical tool for calculating the inner product in the feature space. Let's see how the kernel function works mathematically.

Provided that the chosen kernel function $K(\mathbf{x}_i, \mathbf{x}_j)$ is associated with the corresponding latent function sets $\varphi(\mathbf{x})$, which can map the original data into an H-dimensional feature space (H is usually very large), the following condition holds.

$$K(\mathbf{x}_i, \mathbf{x}_j) = \varphi(\mathbf{x}_i)^t \varphi(\mathbf{x}_j), \quad i, j = 1, 2, 3, \ldots, N \tag{2.23}$$

Now in the feature space, the linear OSH can be constructed using the same procedure as in the linearly separable case (see Section 2.3.1) just by replacing the sample $\mathbf{x}_i$ by $\varphi(\mathbf{x}_i)$. Thus, the optimizing problem in the feature space can be written similarly to (2.17).

$$\text{minimize}: \ L(\mathbf{w}, b, \boldsymbol{\alpha}) = \sum_{i=1}^{N} \alpha_i - \frac{1}{2} \sum_{i,j=1}^{N} y_i y_j \alpha_i \alpha_j \varphi(\mathbf{x}_i)^t \varphi(\mathbf{x}_j)$$

$$\text{subject to}: \ 0 \leq \alpha_i \leq C, \ i = 1, 2, 3, \ldots, N \tag{2.24}$$

By substituting (2.23) into (2.24), we get

$$\text{minimize}: \ L(\mathbf{w}, b, \boldsymbol{\alpha}) = \sum_{i=1}^{N} \alpha_i - \frac{1}{2} \sum_{i,j=1}^{N} y_i y_j \alpha_i \alpha_j K(\mathbf{x}_i, \mathbf{x}_j)$$

$$\text{subject to}: \ 0 \leq \alpha_i \leq C, \ i = 1, 2, 3, \ldots, N \tag{2.25}$$

The decision function can be calculated as

$$f(\mathbf{x}) = sgn\left[\sum_{i=1}^{N} y_i \alpha_i K(\mathbf{x}_i, \mathbf{x}) + b \right] \tag{2.26}$$

where α_i ($0 < \alpha_i \leq C$) is the solution of the QP problem in (2.25), and b can be computed using the equation:

$$b = y_j - \sum_{i=1}^{N} y_i \alpha_i K(\mathbf{x}_i, \mathbf{x}_j), \ j \in \{j \,|\, 0 < \alpha_j \leq C\} \tag{2.27}$$

Finally, to intuitively understand the SVC model, the SVC architecture is shown in Figure 2.3. From the point of view of geometric architecture, it is very similar to the artificial neural networks. Thus SVMs are also sometimes called support vector networks.

2.3.3 Application of SVC to simulated data

A dataset containing 150 samples in 2-D space are simulated and shown in Figure 2.4. It can easily be seen that there exists an apparent region where data of the two classes coexist. The data are linearly insepara-ble. As is known, quadratic discriminant analysis (QDA) is a classical

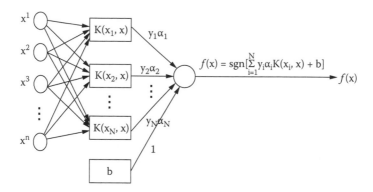

Figure 2.3 The architecture of support vector machines. The input vector $\mathbf{x} = [x^1, x^2, x^3, \ldots, x^n]$. The final decision function is the linear combination of the kernel-based nonlinear transformed outputs plus the bias b with weights vector.

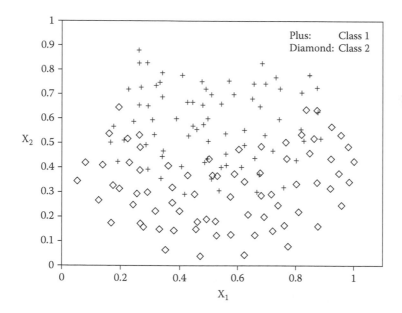

Figure 2.4 The distribution of the simulated data used for classification.

method commonly used in the field of pattern recognition dealing with the nonlinear problem in data. It constructs a nonlinear decision function between different classes. In the present study, QDA is applied as a reference modeling method and the QDA results are compared with those obtained by SVC. The LIBSVM package is employed to build support vector classifiers [13].

For SVC, the choice of kernel function is a key step in constructing the model. In this study, we employ the most frequently used radial basis function as the kernel. Then two parameters, γ (Section 2.2.1) and C (regularizing factor), should be chosen first to construct the model. As known, the appropriate parameter choice has great influence on the stability and predictive performance of the model. So far the most commonly applied technique for determining the parameters of a model is cross-validation (CV) [14–16]. But this technique is computationally intensive and local. Thus, it will usually be difficult to find the best parameter or combination of different parameters in a global and comprehensive manner. However, parameters have to be optimized first in order to obtain a model with good prediction ability. Here, the genetic algorithm (GA) is applied to optimize C and γ because the GA has the ability to globally locate the optimized solution. The GA Toolbox (version 1.2), freely available at the website of the University of Sheffield [17], is employed to conduct parametric optimization. In the GA, the object is currently to minimize the error rate of ten fold cross-validation calculated by the SVC. Finally, the C and γ corresponding to the lowest error rate are taken to train the SVC model using the training set. In ten fold cross-validation, the training set samples are randomly divided into ten groups. Then a model is trained on nine groups and the remaining group is left out for testing each time. This procedure is repeated until all the samples are predicted.

The optimized C and γ obtained by the GA are 60.15 and 0.38, respectively. Then an SVC model is calculated with the training set. An independent test set is used to test the predictive performance of the trained model. The error rate on the training set and the test set together with the total set are all listed in Table 2.1. It can be found that the error rates of SVC are all lower by at least 0.04 than that of QDA, which indicates an obvious improvement. The results further prove that SVC is quite a competent method for discriminating analysis of the nonlinear data. The reason for this may lie in that SVC maximizes the margin between classes in the feature space, which makes the decision boundary far from each class and safe based on the SRM principle. But it should be emphasized here that it does take much more time to use the GA to optimize the parameters for obtaining a good SVC model. On the other hand, QDA takes much less time and

Table 2.1 Error Rates of QDA and SVC on the
Simulated Data for Classification[a]

Method	Fitting Errors	Test Errors	Cross Validation Errors[b]
QDA	0.190	0.240	0.200
SVM	0.140	0.200	0.160

[a] The optimized parameters of SVC are $C = 60.15$ and $\gamma = 0.38$.
[b] The error rates based on 10-fold cross validation on the train set.

is computationally less intensive. In practice, it's worthwhile to spend time to execute the optimization procedure if we care more about the quality of the model. Objectively, they both have their own superiority and disadvantages.

It seems necessary to again elucidate the good generalization performance of SVC from the point of its basic principle. Basically, the original data are nonlinearly mapped into a high-dimensional space and then a linear OSH is constructed with the maximal margin that can ensure the generalizing ability of SVC. Thus it can deal with the nonlinearity in the data and simultaneously minimize the structural risk due to margin maximization. That may be the reason why SVC could perform better in many cases than other statistical or machine-learning methods.

2.4   *Support vector machines for regression*

Support vector machines have been extended to solve the regression problem with the given dataset $D = \{(\mathbf{x}_i, y_i)\}_{i=1}^{N}$ obtained from a latent function where $\mathbf{x}_i$ denotes the sample vector and y_i the corresponding response. N is the total number of samples. Two basic important concepts should be first paid attention to in SVR. One is the ε-band and the other is the ε-insensitive loss function. As with the SVC, the original data are first nonlinearly mapped into a high-dimensional feature space, and then a linear function is fitted to approximate the latent function between $\mathbf{X}$ and $\mathbf{y}$.

2.4.1   *ε-Band and ε-insensitive loss function*

Let's take a single variable function, for example, as shown in Figure 2.5. The ε-band refers to the region between the two broken lines. This region can be found by moving the solid line within a displacement of ε up and down. Here, ε is a predefined positive number.

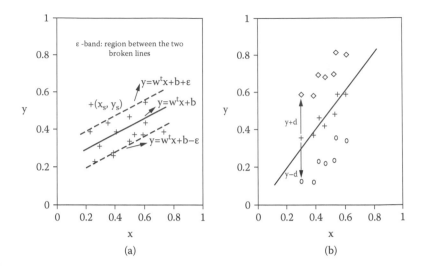

Figure 2.5 Plot (a): the graphical expression of ε-band of the univariate function with the predefined ε.plot; (b): how to transform the regression problem into a classification problem. For each sample $\mathbf{x}_i$ in the training data (plus), the corresponding y_i is added by a positive number d to produce one new sample $(\mathbf{x}_i, y_i^1)$ belonging to Class 1. Similarly, the y_i can also be subtracted by the same d to produce another new sample $(\mathbf{x}_i, y_i^{-1})$ belonging to Class –1. Repeating this procedure, the N samples for regression are doubled and classified into two classes. Then the regression problem is transformed into the binary classification problem. Thus the algorithm for SVC can be applied here to solve the regression problem.

The ε-insensitive loss function has the piecewise form:

$$L(y - f(\mathbf{x}), \varepsilon) = \begin{cases} |y - f(\mathbf{x})| - \varepsilon, & |y - f(\mathbf{x})| \geq \varepsilon \\ 0, & otherwise \end{cases} \tag{2.28}$$

That is, only the data points outside the ε-band (e.g., $(\mathbf{x}_s, y_s)$ in Figure 2.5a) cause loss. This kind of loss function is illustrated in Figure 2.6.

2.4.2   Linear ε-SVR

In order to solve the regression problem, Cortes et al. cleverly transformed the regression problem into the classification problem. Then the regression function can be calculated using the same algorithm as described in SVC. The first step dealing with the regression problem is to convert it into a classification problem, shown in Figure 2.5 (Plot (b)).

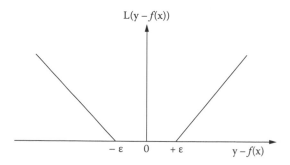

Figure 2.6 An illustration of the ε-insensitive loss function.

Given the training data $D = \{(\mathbf{x}_i, y_i)\}_{i=1}^{N}$, the linear ε-SVR algorithm theoretically aims to solve the optimizing problem, which can be written in the following form with an ε-insensitive loss term:

$$\text{minimize}: \frac{1}{2}|| \mathbf{w}||^2 + \frac{C}{N}\sum_{i=1}^{N} L(y_i - f(\mathbf{x}_i), \varepsilon) \qquad (2.29)$$

where C is a predefined regularizing parameter. The above minimizing problem can further be expressed in the following form with the slack variable $\xi_i^{(*)}$ introduced:

$$\text{minimize}: L(\mathbf{w}, b, \xi^{(*)}) = \frac{1}{2}|| \mathbf{w}||^2 + \frac{C}{N}\sum_{i=1}^{N}(\xi_i + \xi_i^*)$$

$$\text{subject to}: (\mathbf{w}^t\mathbf{x}_i + b) - y_i \leq \varepsilon + \xi_i, \ i = 1, 2, \ldots, N$$

$$\qquad\qquad (2.30)$$

$$y_i - (\mathbf{w}^t\mathbf{x}_i + b) \leq \varepsilon + \xi_i^*, \ i = 1, 2, \ldots, N$$

$$\xi_i^{(*)} \geq 0, \ i = 1, 2, \ldots N$$

With the help of the Lagrange multiplier method and QP algorithm (see Appendix B for more detail), the regression function can be derived as

$$f(\mathbf{x}) = \sum_{i=1}^{N}(\alpha_{i,f}^* - \alpha_{i,f})(\mathbf{x}_i^t\mathbf{x}) + b_f \qquad (2.31)$$

$$b_f = y_j - \sum_{i=1}^{N} (\alpha_{i,f}^* - \alpha_{i,f})(\mathbf{x}_i^t \mathbf{x}_j) + \varepsilon \qquad (2.32)$$

where $\alpha_{i,f}^*$ and $\alpha_{i,f}$ are the optimized Lagrange multipliers, respectively.

2.4.3 Kernel-based ε-SVR

It is well known that nonlinearity is the most common case in real-world datasets. Therefore, it's necessary to extend the linear ε-SVR into nonlinear regression. By introducing the kernel function, the original input was first nonlinearly mapped into the feature space, and the resulting ε-SVR becomes so flexible that it can be used to cope with the complicated nonlinear regression problem in chemistry.

As the derivation procedure of the final decision function is quite similar to that in the linear case, we give here only the ultimate mathematical form:

$$f(\mathbf{x}) = \sum_{i=1}^{N} (\alpha_{i,f}^* - \alpha_{i,f}) K(\mathbf{x}_i, \mathbf{x}) + b_f \qquad (2.33)$$

$$b_f = y_j - \sum_{i=1}^{N} (\alpha_{i,f}^* - \alpha_{i,f}) K(\mathbf{x}_i, \mathbf{x}_j) + \varepsilon \qquad (2.34)$$

where $\alpha_{i,f}^*$ and $\alpha_{i,f}$ are the optimized Lagrange multipliers.

In addition, it should be pointed out that v-SVR ($v \in [0, 1]$) could be treated as a modified version of the original ε-SVR based on the kernel function. In v-SVR, v is a sparsity parameter. After choosing v, the ε value is automatically tuned by the algorithm. Here v is the lower bound of the proportion of support vectors to the total samples and the upper bound of the fraction of errors at the same time.

2.4.4 Application of SVR to simulated data

To specifically illustrate that SVR has the potential to cope with the nonlinearity, an ultraviolet dataset is simulated. It contains the spectra of 60 mixtures composed of four compounds. The spectra together with the concentrations are simulated according to formula (2.35).

$$\begin{cases} \mathbf{X} = \mathbf{SC}^t \\ \mathbf{X} = \mathbf{X} + 0.5\mathbf{X}^2 - \mathbf{X}^3 \\ \mathbf{y} = \mathbf{X}\boldsymbol{\beta} + \mathbf{e} \end{cases} \qquad (2.35)$$

Here, $\mathbf{S}$ is a 30×4 matrix consisting of the pure spectrum of each of the four compounds. $\mathbf{C}$ is a 60×4 mixing matrix containing the concentration of 60 samples (Table 2.2). The concentration for each compound is randomly produced. The resulting $\mathbf{X}$ collected the spectra of the 60 mixtures. The pure spectrum and the mixture spectra are all shown in Figure 2.7. Considering the nonlinearity and inevitable noise in real cases, the nonlinear term (i.e., the quadratic and cubic terms) and the noise ($\sigma = 0.004$) are also added to $\mathbf{X}$. The concentration of the fourth compound is taken as the response value $\mathbf{y}$. Forty out of the sixty samples are randomly chosen as the training set and the remaining twenty are the test set (denoted by the asterisk in Table 2.2).

As is known, partial least squares (PLS) [18–21] is the most widely used calibration method in the field of near-infrared spectroscopy. Here, the performances of both PLS and SVR [2, 22] on the simulated data are explored and compared. Before modeling, the sample matrix $\mathbf{X}$ and the concentration $\mathbf{y}$ are all scaled into the region [0, 1].

For PLS, ten fold cross-validation is utilized to choose the optimal number of latent variables (LVs). The root mean squared errors of cross-validation (RMSECV) as a function of the number of LVs is shown in

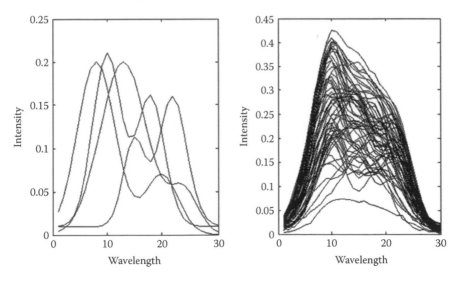

Figure 2.7 The simulated 4 pure spectra (Plot a) and the spectra of the 60 samples with white noise ($\sigma = 0.004$) added (Plot b).

Table 2.2 The Concentrations of the Four Compounds
in the Simulated UV Dataset

NO.	c_1	c_2	c_3	c_4
1	0.5657	0.9214	0.0225	0.4422
2	0.7400	0.9078	0.2609	0.4409
3	0.6915	0.6683	0.7931	0.4814
4	0.2965	0.0789	0.7459	0.4528
5	0.0525	0.8934	0.2700	0.5769
6	0.4727	0.9601	0.0599	0.5075
7	0.2407	0.8774	0.9091	0.4346
8	0.4549	0.8809	0.0179	0.8595
9	0.6501	0.7897	0.6694	0.6157
10	0.5109	0.1756	0.4735	0.1220
11	0.4391	0.6362	0.8745	0.6751
12	0.0208	0.4424	0.7048	0.5099
13	0.9753	0.1074	0.4777	0.7757
14	0.4504	0.1314	0.3505	0.3542
15	0.4017	0.2685	0.0966	0.9114
16	0.5835	0.4053	0.9836	0.1370
17	0.1308	0.5498	0.3813	0.2273
18	0.4500	0.7789	0.2542	0.6168
19	0.7422	0.2594	0.7904	0.8245
20	0.0421	0.4651	0.4990	0.6553
21	0.1779	0.1718	0.6382	0.8919
22	0.3127	0.2658	0.0210	0.8143
23	0.7659	0.7430	0.1497	0.7531
24	0.4678	0.6592	0.6211	0.1433
25	0.5578	0.0766	0.1792	0.5776
26	0.1663	0.7410	0.7674	0.9662
27	0.7053	0.3715	0.5696	0.5696
28	0.5431	0.2417	0.7763	0.5161
29	0.6212	0.8690	0.0093	0.7661
30	0.0744	0.2327	0.1080	0.9115
31	0.5306	0.6271	0.3381	0.8873
32	0.8746	0.6413	0.0518	0.8824
33	0.0979	0.5208	0.3760	0.6813
34	0.2903	0.8737	0.9073	0.2327
35	0.9136	0.1651	0.4943	0.7467
36	0.0833	0.5848	0.2900	0.5901
37	0.1047	0.5334	0.0262	0.4668

Table 2.2 The Concentrations of the Four Compounds
in the Simulated UV Dataset (Continued)

NO.	c_1	c_2	c_3	c_4
38	0.0548	0.5026	0.5671	0.0291
39	0.4734	0.6079	0.9063	0.4022
40	0.5176	0.1942	0.4162	0.0717
41*	0.3636	0.1690	0.7846	0.4263
42*	0.1064	0.0335	0.7697	0.3478
43*	0.5245	0.6052	0.9714	0.0265
44*	0.9423	0.1422	0.1787	0.4900
45*	0.3766	0.2595	0.8397	0.0284
46*	0.4216	0.7994	0.7938	0.6937
47*	0.3080	0.9669	0.0584	0.1227
48*	0.2388	0.8193	0.3193	0.8909
49*	0.4750	0.0428	0.1750	0.9163
50*	0.9378	0.6055	0.0385	0.3354
51*	0.0622	0.9020	0.1227	0.8151
52*	0.0223	0.1799	0.1649	0.1221
53*	0.5607	0.4189	0.7959	0.7982
54*	0.4692	0.6208	0.0051	0.3348
55*	0.1742	0.5462	0.2037	0.8420
56*	0.9613	0.0680	0.8002	0.1954
57*	0.6535	0.8603	0.5352	0.2860
58*	0.5511	0.9023	0.7720	0.7245
59*	0.8989	0.5951	0.5034	0.3346
60*	0.3936	0.0260	0.5781	0.0012

* The samples used for test.

Figure 2.8. From Figure 2.8, it can be observed that the RMSECV value begins to increase after 10 LVs. So, 10 PLS components are chosen to build the PLS model.

v-SVR is applied here to construct the SVR model. The RBF is used as the kernel. The γ of the RBF kernel is set to the default value in the LIBSVM software. There are two parameters to be predefined before training. One is the regularizing factor C; the other is the sparsity parameter v. The genetic algorithm is again used to optimize these two parameters globally. The optimized values of C and v by GA are 8.09 and 0.9881, respectively (Table 2.3). Finally, the SVR model is built with the optimized parameters using the training set.

Both the results obtained by PLS and SVR are presented in Table 2.3. The prediction errors of SVR for both the training set and test set are lowered by 22.1 and 9.5% compared to those of PLS, which indicates an

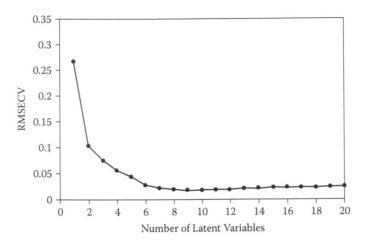

Figure 2.8 Root mean squared errors of cross-validation (RMSECV) values of ten fold cross-validation resulted from PLS for the simulated UV data.

Table 2.3 The Performance Parameters of PLS and SVR for the Simulated UV Data[a]

Method	RMSEF (Train)	RMSEP (Test)	RMSECV (Train)	R^2
PLS	0.0103	0.0213	0.0167	0.9973
SVMs	0.0073	0.0199	0.0186	0.9979

[a] The GA-optimized parameters of SVR are C=8.09 and v=0.9881. Notice that the original y-value are scaled into the region [0, 1] when building PLS and SVR model. But the results in this table are reported after transforming the scaled y-value into the original unit. RMSEF: root mean squared errors of fitting. RMSEP: root mean squared errors of prediction. RMSECV: root mean squared errors of cross validation.

obvious improvement. The squared correlation coefficients of PLS and SVR on the whole dataset are 0.9737 and 0.9835, respectively. The predicted and experimental values are shown in Figure 2.9. From Table 2.3 and Figure 2.9, it can be seen that SVR gives both better fitting and predictive performance.

The results from the simulation show that SVR can predict concentration more accurately. The reason may be that some nonlinearity component has been manually added to the simulated dataset, such that the linear PLS can't work well. But the kernel-based SVRs have the ability to handle the nonlinear data and to model both the linear and nonlinear relationship between the spectra and the concentration. The simulated

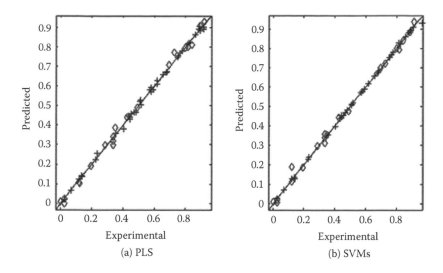

Figure 2.9 The predicted concentration versus experimental values for UV data:
(a) PLS, (b) SVMs.

study clearly shows that SVR is indeed a good alternative for regression analysis for the data of some nonlinearity. In all, SVR is capable of dealing with both linear and nonlinear relationships. Thus, we may conclude that SVR is more powerful in capturing the latent data structure and modeling the unknown nonlinearity. From this point of view, SVR seems to be a promising method for modeling real-world data of inherent nonlinearity.

2.5   Parametric optimization for support vector machines

Generally after a model either for classification or for regression is theoretically developed, the next step is to design an algorithm to implement the theory. The algorithm should have its own tuning parameters telling the algorithm to what extent the data should be learned or how the algorithm should be stopped or something else. As is known, the choice of tuning parameters is a deterministic factor that underlies the predictive performance of the trained model. The most frequently mentioned issues concerning model building are the so-called overfitting and underfitting. Loosely speaking, the former refers to the situation where the data are so exhausted that the unnecessary part of the data, for example, noise, is learned by the model. In this case, the established model usually cannot predict the future sample with good accuracy. On the contrary, the latter refers to the fact that the data are not learned sufficiently and hence the model misses, more or less, some necessary information in the data. The resulting model would not yield good performance yet. It is therefore very important for us to optimize the tuning parameters carefully before constructing the final model, especially when a nonlinear algorithm is employed.

In the context of support vector machines, the tuning parameters could be partitioned into two categories. The first is associated with the algorithm of the support vector machines used. For C-SVC, v-SVC, ε-SVR, and v-SVR, the parameters (i.e., C, v, and ε) fall into the first category. The second class of tuning parameters is related to the user-chosen kernel function, for example, the kernel width γ of the Gaussian kernel (see Section 2.2.1).

To give an intuitive understanding of how the tuning parameters influence the predictive performance of the constructed SVM regression model, let's first see some results on the same simulated data in Section 2.4.4. We choose ε-SVR with a radial basis function as the kernel to build the SVR model. Note that there are three tuning parameters for ε-SVR: C, ε, and the kernel width γ.

We first tune the value γ in the range of [0.001, 0.01] with C and ε fixed at 200 and 0.001, respectively. For each γ, a SVR model is first built on the training set of 40 samples and the root mean squared errors of fitting (RMSEF) are computed. Then, the root mean squared errors of prediction (RMSEP) are also calculated using the test set. The values of RMSEF and RMSEP are shown in Figure 2.10. Obviously, the RMSEF value decreases consistently with increased γ. However, the changing trend of RMSEP is different. It first decreases quickly, then achieves the minimum approximately at $\log_{10}\gamma = -3.5$, and finally begins to rise. The reason for the observation may be that overfitting to some extent occurs when $\log_{10}\gamma > -3.5$, which makes the established SVR model become overfitted and degrade the predictive performance. So, it may be concluded that the model builder should carefully adjust the parameters in order to obtain a well-generalized model.

By analogy, we also tune the value ε in the range of [0.0001, 0.01] with C and γ fixed at 200 and 0.001, respectively. Both RMSEF and RMSEP values against ε are plotted in Figure 2.11. A similar phenomenon to that discussed in the last case can be observed. With decreased ε, the RMSEP value also first decreases. Then it begins to increase, indicating, in our opinion, that the SVR model is to some extent overfitted.

Finally, the influence of the penalizing factor C on the predictive ability of the SVR model is also investigated. The range of C is set to [5, 200], whereas C and γ are fixed at 200 and 0.001, respectively. In this setting, both RMSEF and RMSEP values are computed and shown in Figure 2.12. Obviously, C is also an important factor that underlies the performance of the SVR model.

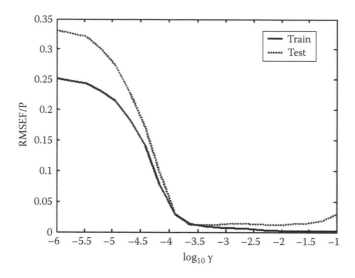

Figure 2.10 The RMSEF and RMSEP as a function of γ with C and ε set to 200 and 0.001, respectively.

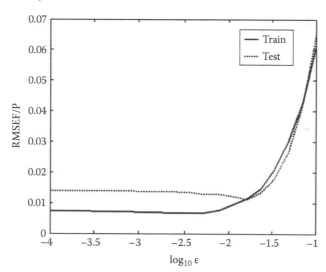

Figure 2.11 The RMSEF and RMSEP as a function of ε with C and γ set to 200 and 0.001, respectively.

Based on the findings of the case study, it can be strongly claimed that the optimization of SVM tuning parameters is the key step for obtaining a high-quality model. SVMs, in our opinion, are theoretically a good machine-learning method because of their intrinsic distinctions, such as margin maximization. But in fact, through our experience, whether the

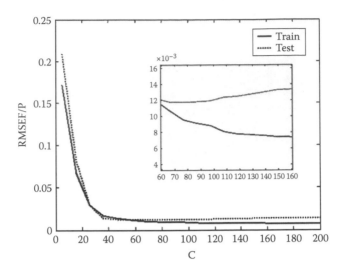

Figure 2.12 The RMSEF and RMSEP as a function of C with ε and γ set to 0.001 and 0.001, respectively.

SVM can be applied successfully is heavily dependent on the choice of tuning parameters. Notice also that so far we have only examined the influence of a single parameter with other parameters set to fixed values. The influence of the combination of different parameters on the SVM model has not yet been considered.

To simultaneously optimize several parameters, we have to resort to techniques having the potential to locate an optimal set of tuning parameters. Such techniques include, but are not limited to, grid search (GS), orthogonal design, uniform design, genetic algorithms, simulated annealing, differential evolution, and so on. But how can one perform optimization tasks using these methods? First, it should be made clear that any optimization task consists of at least two elements: (1) the decision variables and their feasible region and (2) the objective function to be optimized (either maximized or minimized). Take ε-SVR and a training set of p variables as an example; one possible solution is that (1) set $C \in [5,200]$, $ε \in [0.001, 0.01]$ and $γ \in [0.001, 1]$ and (2) the objective function are set to minimize the cross-validated prediction errors of SVM. The readers are referred to both Sections 2.3.3 and 2.4.4 for some details. Some applications related to parametric optimization can also be found in Chapters 5 and 6. So we do not give a concrete example here. Let's now turn to another topic, "variable selection for support vector machines," which is also very important for improving the predictive performance of a SVM model.

2.6   Variable selection for support vector machines

In the fields of chemistry and biotechnology, routinely produced data by modern analytical instruments are in most cases composed of hundreds of thousands of variables, especially in the area of near-infrared spectroscopy (NIR) [23–27] and OMICS studies, such as genomics [28–34], proteomics [33,35–39], and metabolomics [40–44]. Let's see some examples first. (1) A near-infrared spectrum contains the information measured on over 1,000 wavelengths, (2) a microarray experiment on only one gene chip can produce the expression profile of over 1,000,000 genes, (3) the data of a protein mixture sample subjected to a MALDI-TOF experiment can be of over 10,000 dimensions, and (4) in a metabolomics study, the NMR data or LC/MS or GC/MS are also high-dimensional.

Although SVMs have been applied successfully in many fields, variable selection has proved to be effective for improving performance [45–47]. So far, there is not much literature addressing variable selection for support vector machines. One possible reason may be that SVMs are a nonlinear and kernel-based modeling method, making it difficult for us to understand the role or importance level of the variable. Based on model population analysis (MPA) [48,49], we have specially designed a strategy, called margin influence analysis (MIA), for variable selection of support vector machines [50]. This method is briefly introduced in Chapter 6 with applications to traditional Chinese medicines (TCMs). This section is only aimed at putting forward the viewpoint that variable selection has the potential to (greatly) improve the performance of SVM. One typical example could be found in the work of Guyon et al. [45] where recursive feature elimination (RFE) is employed to remove unimportant variables step by step. In later chapters we describe how to perform variable selection for support vector machines.

2.7   Related materials and comments

So far, we have given an overall picture of both the theory and algorithm of support vector machines. Moreover, the two key factors underlying the performance of SVMs (i.e., the optimization of tuning parameters and variable selection) have also been touched on. For understanding the mechanism of SVMs, all the above-introduced materials are personally enough. However, we are planning to introduce some additional considerations related to the practical usage of SVMs, aiming at giving a more comprehensive description. Here, two important concepts are first briefly described. One is the VC (Vapnik–Chervonenkis) dimension and the other is quadratic programming. Finally, some personal opinions on dimension increasing versus dimension reducing are also discussed in some detail.

2.7.1   VC dimension

In general, the VC dimension is a rather complex but important concept in statistical learning theory. It is related not only to a fast rate of convergence of a learning machine but also to the generalization ability of a learning machine. From a practical point of view, we should focus on the influence of VC dimension on the generalization ability of a learning machine. Figure 2.13 shows how the risk bound (generalization ability), confidence interval, and empirical risk of learning machines change with the increasing VC dimension h [7]. From this plot, it is clear that the risk bound will achieve a minimum when VC dimension increases from a low value to a large one, indicating that the choice of VC dimension is very important for building a high-performance model. This is somewhat like selecting an optimal number of latent variables in principal component regression and partial least squares. Notice that VC dimension cannot be computed for all functions. It can only be computed in some special cases. Taking the set of linear indicator functions (2.36) as an example [7], the VC

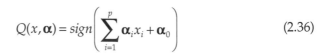

$$Q(x, \boldsymbol{\alpha}) = sign\left(\sum_{i=1}^{p} \alpha_i x_i + \alpha_0\right) \tag{2.36}$$

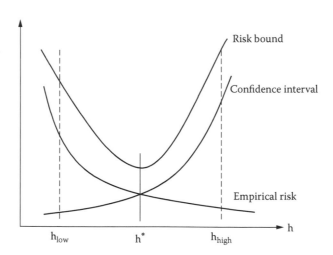

Figure 2.13 The influence of VC dimension h on the risk bound, confidence interval, and empirical risk of learning machines.

dimension of this function set is $p + 1$. Suppose the training data are of m samples and p variables ($p \gg m$; this is commonly encountered in chemistry and biology), we could expect, from Figure 2.13 that the linear learning machine trained on these data should be of high risk bound and hence have poor prediction ability on the new data. Therefore, we should reduce the VC dimension in order to achieve better performance. Personally, variable selection is an effective solution to reduce the VC dimension and can, in most cases, significantly improve the prediction ability of the model. The variable selection of SVMs is addressed in Chapters 7 and 8.

2.7.2   Kernel functions and quadratic programming

As discussed before, one important element of SVMs is the kernel function. By kernel transformation, a symmetric and positive-definite kernel matrix can be obtained. Thus, with the help of the kernel function, the complex optimization procedure in SVMs can be transferred into a convex quadratic programming problem, which has a global minimum and could guarantee that the margin of the SVM classifier is maximized when the other tuning parameters are fixed. It is also worthwhile noting that there are many standard programs to solve the QP problem so far. For example, in MATLAB® (version 7.6.0), the function *quanprog.m* can be called to perform the QP task in an easy way.

2.7.3   Dimension increasing versus dimension reducing

Here, we have a general problem in the community of statistical modeling. It is, in our opinion, more like a philosophy problem. We are not sure our viewpoint is right, but we do deem it worth considering.

We start with ordinary least squares (OLS). Given a dataset $(\mathbf{X}, \mathbf{y})$ of m samples and p variables, a linear model of p dimension can be established and then could be used for making predictions. If $p > m$ or the variables are highly collinear, ridge regression (RR) may be employed instead of OLS. Here, for convenience, we say the OLS or RR model is constructed in the original feature space, shown in Figure 2.14.

In addition to OLS and RR, the latent variable-based modeling method, for example, PCR and PLS, may be the most widely used technique, especially in chemistry. Part of the reason is that PCR and PLS have proved to be very effective for high-dimension data in the community of chemometrics and the industrial field related to process monitoring and the like. Mathematically, the key point of PCR and PLS lies in the fact that they work by regressing the response variable $\mathbf{y}$ on only a few latent variables (reduced feature space), also shown in Figure 2.14. Without question, such models may be parsimonious with the model built in the original space.

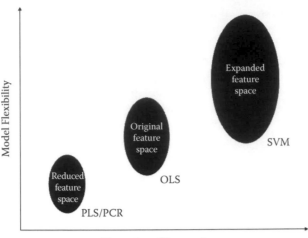

Figure 2.14 Model flexibility versus the dimension of feature space.

In contrast, it can be learned from the theory of support vector machines that the SVM model is built in most cases in a high-dimensional feature space (expanded feature space), shown in Figure 2.14. The SVMs model is thus very flexible. Surprisingly, SVM can also perform very well, which is at least empirically demonstrated in a large amount of the literature.

Naturally based on the above discussion, one may ask, "Should we build a model in the reduced feature space or in the expanded feature space?" A practitioner may answer, "We only choose the one with better performance, despite its being built in reduced feature space (M1) or expanded space (M2)." But if two models M1 and M2 are of nearly the same prediction ability (for a given datum), then what's the choice? Should we take Occam's razor as our standard? Maybe there exists a variety of viewpoints or answers. As chemists or biologists, our opinions could be deconvoluted into two points. First, we advocate that the prior information is very important and hence should be elegantly incorporated into the modeling procedure. Taking the quantitative analysis of mixture in chemistry as an example, we of course should first choose the linear model to correlate the concentration with the spectral data because of the known Lambert–Beer law (in the linear range) and other similar additive rules. However, if the model is unknown, that is, there is no prior information at hand, we secondly prefer the simpler one if it can work well. If the data are so complex that they cannot be described well by a linear model, we resort to the nonlinear method as with the two examples shown in this chapter. In a few words, the simplest model that can work well is our choice. On the whole, model comparison and selection are very fundamental and important in both statistics and chemometrics, and to date there exist a number of issues. It is expected that the (empirical) Bayesian method would be a promising theory for solving the model comparison and selection problem.

Appendix A: Computation of slack variable-based SVMs

The Lagrange function of this optimizing Problem (2.21) is

$$L(\mathbf{w},b,\xi,\boldsymbol{\alpha},\mathbf{r}) = \frac{1}{2}\mathbf{w}^t\mathbf{w} + c\sum_{i=1}^{N}\xi_i - \sum_{i=1}^{N}\alpha_i[y_i(\mathbf{w}^t\mathbf{x}_i + b) - 1 + \xi_i] - \sum_{i=1}^{N}r_i\xi_i \quad (A1)$$

where α_i and r_i are the Lagrange multipliers. Setting the respective derivatives against $\mathbf{w}$, ξ, and b to zero, we get

$$\frac{\partial L(\mathbf{w},b,\xi,\boldsymbol{\alpha},\mathbf{r})}{\partial \mathbf{w}} = \mathbf{w} - \sum_{i=1}^{N}y_i\alpha_i\mathbf{x}_i = 0 \quad (A2)$$

$$\frac{\partial L(\mathbf{w},b,\xi,\boldsymbol{\alpha},\mathbf{r})}{\partial \xi} = C - \alpha_i - r_i = 0 \quad (A3)$$

$$\frac{\partial L(\mathbf{w},b,\xi,\boldsymbol{\alpha},\mathbf{r})}{\partial b} = \sum_{i=1}^{N}y_i\alpha_i = 0 \quad (A4)$$

By inserting (A2)–(A4) into (A1), we obtain the dual objective function of (A1):

$$L(\mathbf{w},b,\xi,\boldsymbol{\alpha},\mathbf{r}) = \sum_{i=1}^{N}\alpha_i - \frac{1}{2}\sum_{i,j=1}^{N}y_iy_j\alpha_i\alpha_j\mathbf{x}_i^t\mathbf{x}_j \quad (A5)$$

In addition, the Karush–Kuhn–Tucker condition of this optimizing problem includes the constraint below:

$$\alpha_i[y_i(\mathbf{x}_i^t\mathbf{w} + b) - 1 + \xi_i] = 0 \quad (A6)$$

The minimization of (A5) is also a QP problem. The solution for $\mathbf{w}$ has the form

$$\mathbf{w} = \sum_{i}^{N}\alpha_iy_i\mathbf{x}_i \quad (A7)$$

And from (A6), it can easily be seen that any of the margin points with $\alpha_i > 0$ and $\xi_i = 0$ can be used to calculate b.

$$b = \frac{1}{y_i} - \mathbf{x}_i^t \mathbf{w} \tag{A8}$$

But usually we use the average of all the solutions for numerical stability. Given the solutions **w** and b, the decision function can be calculated as

$$f(\mathbf{x}) = sgn(\mathbf{w}^t \mathbf{x} + b) = sgn\left[\left(\sum_{i=1}^{N} y_i \alpha_i \mathbf{x}_i\right)^t \mathbf{x} + b\right] \tag{A9}$$

Appendix B: Computation of linear ε-SVR

With the introduced Lagrange multipliers α_i and α_i^*, the Lagrange function of the optimizing problem in (2.30) is

$$L(\mathbf{w}, b, \boldsymbol{\alpha}, \xi^{(*)}) = \frac{1}{2}||\mathbf{w}||^2 + \frac{C}{N}\sum_{i=1}^{N}(\xi_i + \xi_i^*) - \sum_{i=1}^{N}(\eta_i\xi_i + \eta_i^*\xi_i^*)$$

$$-\sum_{i=1}^{N}\alpha_i(\varepsilon + \xi_i + y_i - \mathbf{w}^t\mathbf{x}_i - b) - \sum_{i=1}^{N}\alpha_i^*(\varepsilon + \xi_i^* - y_i + \mathbf{w}^t\mathbf{x}_i + b) \tag{B1}$$

Deriving $L(\mathbf{w}, b, \boldsymbol{\alpha}, \xi^{(*)})$ against w, b, and $\xi^{(*)}$ we obtain the following equations:

$$\frac{\partial L(\mathbf{w}, b, \boldsymbol{\alpha}, \xi^{(*)})}{\partial \mathbf{w}} = \mathbf{w} - \sum_{i=1}^{N}(\alpha_i^* - \alpha_i)\mathbf{x}_i = 0 \tag{B2}$$

$$\frac{\partial L(\mathbf{w}, b, \boldsymbol{\alpha}, \xi^{(*)})}{\partial b} = \sum_{i=1}^{N}(\alpha_i - \alpha_i^*) = 0 \tag{B3}$$

$$\frac{\partial L(\mathbf{w}, b, \boldsymbol{\alpha}, \xi^{(*)})}{\partial \xi^{(*)}} = \frac{C}{N} - \alpha_i^* - \eta_i^* = 0 \tag{B4}$$

Inserting Equations (B2), (B3), and (B4) into (B1), we can get the dual form of the optimizing problem:

$$\text{minimize}: \frac{1}{2} \sum_{i,j=1}^{N} (\alpha_i^* - \alpha_i)(\alpha_j^* - \alpha_j)(x_i^t x_j) + \varepsilon \sum_{i=1}^{N} (\alpha_i^* + \alpha_i) - \sum_{i=1}^{N} y_i(\alpha_i^* - \alpha_i) \quad \text{(B5)}$$

$$\text{subject to}: \sum_{i=1}^{N} (\alpha_i - \alpha_i^*) = 0 \quad \text{(B6)}$$

$$0 \le \alpha_i, \alpha_i^* \le \frac{C}{N}, \; i = 1, 2, \ldots, N \quad \text{(B7)}$$

By solving the above dual problem, the final decision function can be derived:

$$f(x) = \sum_{i=1}^{N} (\alpha_{i,f}^* - \alpha_{i,f})(x_i^t x) + b_f \quad \text{(B8)}$$

$$b_f = y_j - \sum_{i=1}^{N} (\alpha_{i,f}^* - \alpha_{i,f})(x_i^t x_j) + \varepsilon \quad \text{(B9)}$$

where $\alpha_{i,f}^*$ and $\alpha_{i,f}$ are the optimized solutions of (B5).

References

1. Minsk, M.L. and Papert, S.A. 2009. *Perceptrons.* Cambridge, MA: MIT Press.
2. Li, H.-D., Liang, Y.-Z., Xu, Q.-S. Support vector machines and its applications in chemistry. *Chemometr. Intell. Lab.*, 95:188–198.
3. Walczak, B. and Massart, D.L. 1996. Application of radial basis functions— Partial least squares to non-linear pattern recognition problems: Diagnosis of process faults. *Anal. Chim. Acta*, 331(3):187–193.
4. Walczak, B. and Massart, D.L. 1996. The radial basis functions—Partial least squares approach as a flexible non-linear regression technique. *Anal. Chim. Acta*, 331(3):177–185.
5. Baudat, G. and Anouar, F. 2000. Generalized discriminant analysis using a kernel approach. *Neural Comput.*, 12:2385–2404.
6. Mika, S., Ratsch, G., Weston, J., Scholkopf, B., and Muller, K. 1999. Fisher discriminant analysis with kernels. In: *IEEE Neural Networks for Signal Processing Workshop: 1999*, pp. 41–48.

7. Vapnik, V. 1999. *The Nature of Statistical Learning Theory*, second edition, New York: Springer.
8. Vapnik, V. 1998. *Statistical Learning Theory*. New York: Wiley.
9. Hastie, T., Tibshirani, R., and Friedman, J.H. 2001. *The Elements of Statistical Learning: Data Mining Inference and Prediction*. New York: Springer.
10. Shieh, M.-D. and Yang, C.-C. 2008. Multiclass SVM-RFE for product form feature selection. *Expert Syst. Appl.*, 35(1–2):531–541.
11. Angulo, C., Parra, X., and Catal, A. 2003. K-SVCR. A support vector machine for multi-class classification. *Neurocomputing*, 55(1-2):57–77.
12. Peng, S., Zeng, X., Li, X., Peng, X., and Chen, L. 2009. Multi-class cancer classification through gene expression profiles: MicroRNA versus mRNA. *J.Genet. Genom.*, 36(7):409–416.
13. http://www.csie.ntu.edu.tw/~cjlin/libsvm.
14. Stone, M. 1974. Cross-validatory choice and assessment of statistical predictions. *J. Roy. Statist. Soc. B*, 36:111–147.
15. Xu, Q.-S. and Liang, Y.-Z. 2001. Monte Carlo cross validation. *Chemometr. Intell. Lab.*, 56(1):1–11.
16. Filzmoser, P., Liebmann, B., and Varmuza, K. 2009. Repeated double cross validation. *J. Chemometr.*, 23(4):160–171.
17. http://www.shef.ac.uk/acse/research/ecrg/getgat.html.
18. Xu, Q.-S., Liang, Y.-Z.L., and Shen, H.-L. 2001. Generalized PLS regression. *J. Chemometr.* 15(3):135–148.
19. Wold, S., Sjöström, M., and Eriksson, L. 2001. PLS-regression: A basic tool of chemometrics. *Chemometr. Intell. Lab.* 58(2):109–130.
20. De Jong, S. 1993. SIMPLS: An alternative approach to partial least squares regression. *Chemometr. Intell. Lab.* 18(3):251–263.
21. Geladi, P. and Kowalski, B.R. 1986. Partial least-squares regression: A tutorial. *Anal. Chim. Acta*, 185:1–17.
22. Smola, A.J. and Scholkopf, B. 2004. A tutorial on support vector regression. *Statist. Comput.*, 14:199–222.
23. Zhou, Q., Xi, Y., He, H., and Frost, R.L. 2008. Application of near infrared spectroscopy for the determination of adsorbed p-nitrophenol on HDTMA organoclay–Implications for the removal of organic pollutants from water. *Spectrochim. Acta A*, 69(3):835–841.
24. Moros, J., Llorca, I., Cervera, M.L., Pastor, A., Garrigues, S., and de la Guardia, M. 2008. Chemometric determination of arsenic and lead in untreated powdered red paprika by diffuse reflectance near-infrared spectroscopy. *Anal. Chim. Acta*, 613(2):196–206.
25. Luo, X., Yu, X., Wu, X., Cheng, Y., and Qu, H. 2008. Rapid determination of *Paeoniae radix* using near infrared spectroscopy. *Microchem. J.*, 90(1):8–12.
26. Kim, J., Hwang, J., and Chung, H. 2008. Comparison of near-infrared and Raman spectroscopy for on-line monitoring of etchant solutions directly through a Teflon tube. *Anal. Chim. Acta*, 629(1–2):119–127.
27. Watson, C.A. 1977. Near infrared reflectance spectrophotometric analysis of agricultural products. *Anal. Chem.*, 49(9):835A–840A.
28. Cawley, G.C. and Talbot, N.L.C. 2006. Gene selection in cancer classification using sparse logistic regression with Bayesian regularization. *Bioinformatics*, 22(19):2348–2355.
29. Dettling, M. 2004. Bag boosting for tumor classification with gene expression data. *Bioinformatics*, 20(18):3583–3593.

30. Hastie, T., Tibshirani, R., Botstein, D., and Brown, P. 2001. Supervised harvesting of expression trees. *Genome Biol.*, 2:research0003.0001–0003.0012.
31. Ma, S. and Huang, J. 2005. Regularized ROC method for disease classification and biomarker selection with microarray data. *Bioinformatics*, 21(24):4356–4362.
32. Nguyen, D. and Rocke, D.M. 2002. Tumor classification by partial least squares using microarray gene expression data. *Bioinformatics*, 18:39–50.
33. Tyers, M. and Mann, M. 2003. From genomics to proteomics. *Nature*, 422(6928):193–197.
34. Zuber, V. and Strimmer, K. 2009. Gene ranking and biomarker discovery under correlation. *Bioinformatics*, btp460.
35. Aebersold, R. and Mann, M. 2003. Mass spectrometry-based proteomics. *Nature*, 422(6928):198–207.
36. Alexandrov, T., Decker, J., Mertens, B., Deelder, A.M., Tollenaar, R.A.E.M., Maass, P., and Thiele, H. 2009. Biomarker discovery in MALDI-TOF serum protein profiles using discrete wavelet transformation. *Bioinformatics*, 25(5):643–649.
37. Jimmy, K.E., Ashley, L.M., and John, R. 1994. An approach to correlate tandem mass spectral data of peptides with amino acid sequences in a protein database. *J. Amer. Sot..Mass Spectrom.*, 5:976–989.
38. Kuster, B., Schirle, M., Mallick, P., and Aebersold, R. 2005. Scoring proteomes with proteotypic peptide probes. *Nature Rev. Mol. Cell Biol.*, 6(7):577–583.
39. Tang, H., Arnold, R.J., Alves, P., Xun, Z., Clemmer, D.E., Novotny, M.V., Reilly, J.P., and Radivojac, P. 2006. A computational approach toward label-free protein quantification using predicted peptide detectability. *Bioinformatics*, 22(14):e481–488.
40. Gavaghan, C.L., Holmes, E., Lenz, E., Wilson, I.D., and Nicholson, J.K. 2000. An NMR-based metabonomic approach to investigate the biochemical consequences of genetic strain differences: Application to the C57BL10J and Alpk:ApfCD mouse. *FEBS Lett.*, 484(3):169–174.
41. Kamleh, M.A., Hobani, Y., Dow, J.A.T., and Watson, D.G. 2008. Metabolomic profiling of Drosophila using liquid chromatography Fourier transform mass spectrometry. *FEBS Lett.*, 582(19):2916–2922.
42. Madsen, R., Lundstedt, T., and Trygg, J. 2010. Chemometrics in metabolomics—A review in human disease diagnosis. *Anal. Chim. Acta*, 659(1-2):23–33.
43. Yi, L.-Z., He, J., Liang, Y.-Z., Yuan, D.-L., and Chau, F.-T. 2006. Plasma fatty acid metabolic profiling and biomarkers of type 2 diabetes mellitus based on GC/MS and PLS-LDA. *FEBS Lett.*, 580(30):6837–6845.
44. Yuan, D., Liang, Y., Yi, L., Xu, Q., and Kvalheim, O.M. 2008. Uncorrelated linear discriminant analysis (ULDA): A powerful tool for exploration of metabolomics data. *Chemometr. Intell. Lab.*, 93(1):70–79.
45. Guyon, I., Weston, J., Barnhill, S., and Vapnik, V. 2002. Gene selection for cancer classification using support vector machines. *Mach. Learn.*, 46(1):389–422.
46. Bierman, S. and Steel, S. 2009. Variable selection for support vector machines. *Commun. Statist. Simul. Comput.*, 38(8):1640–1658.
47. Zhang, H.H. 2006. Variable selection for support vector machines via smoothing spline ANOVA. *Statist.Sinica*, 16(2):659–674.
48. Li, H.-D., Liang, Y.-Z., Xu, Q.-S., and Cao, D.-S. 2010. Model population analysis for variable selection. *J. Chemometr.*, 24(7–8):418–423.

49. Li, H.-D., Zeng, M.-M., Tan, B.-B., Liang, Y.-Z., Xu, Q.-S., and Cao, D.-S. 2010. Recipe for revealing informative metabolites based on model population analysis. *Metabolomics*, 6(3) 353–361.
50. Li, H.-D., Liang, Y.-Z., Xu, Q.-S., and Cao, D.-S. 2010. Recipe for uncovering predictive genes using support vector machines based on model population analysis (accepted). IEEE/ACM *Trans. Comput. Biol. Bioinf.*

chapter three

Kernel methods

Contents

3.1 Introduction

As we discussed in the last chapter, the kernel function used in SVM offers an alternative solution by projecting the data into a high-dimensional feature space to increase computational power. The linearly inseparable data in the original low-dimensional space could become linearly separable after being projected into the high-dimensional feature space, which is called dimensional superiority in SVMs [1]. What interests us more is that the kernel function can perform nonlinear mapping to the high-dimensional feature space in an implicit manner without increasing computational cost. In fact, the kernel function used in SVMs can be further extended into a class of methods that could solve the nonlinear problems in chemistry and biotechnology by using a simple linear transformation method. It might be regarded as a general protocol to deal with nonlinear data in chemistry and biotechnology. This is the main topic of this chapter.

In chemistry and biotechnology, several multivariate linear techniques, such as multiple linear regression (MLR), ridge regression (RR), principal component regression (PCR), and partial least squares regression (PLSR) are very popular. They are usually used to construct a mathematical model that relates the multiwavelength spectral response to analyte concentration, molecular descriptors to biological activity, and multivariate process conditions/states to final product attributes and so on. Such a model can in general be used to efficiently predict the properties of new samples.

In practice, however, some chemical systems and problems in biotechnology are probably nonlinear ones. In some practical situations dealing with complex chemical systems, linear methods may be inappropriate for describing the underlying data structure because such systems may exhibit significant nonlinear characteristics, whereas the linear models assume that the chemical data are linear. Of course, if some chemical datasets indeed exhibit linear relationships or even nearly linear relationships, the data processing will be greatly simplified, and the results obtained will be more reliable. Modeling nonlinear chemical systems is a very difficult task, because the analytical form of the relation between measured properties and prediction properties of interest is usually unknown. How to effectively cope with such an issue has been a major concern for chemists and biologists for a long time.

To begin, several nonlinear regression techniques are initially proposed to approximate the properties of interest in chemistry. We can generally divide these methods into four groups: (1) methods based on smoothers, for example, alternating conditional expectations (ACE) [2] and smart smooth multiple additive regression technique (SMART) [2]; (2) methods based on splines, for example, classification and regression tree (CART) [3] and multivariate adaptive regression splines (MARS) [4]; (3) nonlinear PLS methods [5–10], for example, spline-PLS, RBF-PLS, and poly-PLS; (4) neural networks, for example, back-propagation network (BPN) and radial basis function network (RBFN).

An excellent tutorial about the first two groups of approaches can be found in References [2,3,11], and about neural networks (NN) in [12–14]. These nonlinear techniques for the first time made it possible to cope with many nonlinear chemical problems to a large extent. However, they still suffer from some practical difficulties. For example, the approaches in the first group often need too many terms and too many adjustable parameters. Building such a model is usually a time-consuming and tedious task. The well-known nonlinear algorithms such as CART, MARS, and NN are based on gradient descent or greedy heuristics and so suffer from local minima. They also frequently suffer from overfitting because their statistical behaviors were not well understood [11]. Moreover, the learned nonlinear functions are usually not robust; that is, they have a low reliability of the prediction results.

The development of kernel methods [15–17], in particular support vector machines (SVMs) [1,18–21], has overcome the computational and statistical difficulties. The methods enable researchers to analyze nonlinear relations with the efficiency that had previously been reserved for linear algorithms. Furthermore, advances in their statistical analysis made it possible to do so in high–dimensional feature spaces while avoiding the dangers of overfitting, because they are strictly established based on statistical learning theory (SLT), a widely recognized theory of statistical science. From all points of view, computational, statistical, and conceptual, kernel-based nonlinear algorithms are very efficient and as well founded as linear ones. The problems of local minima and overfitting that were typical of NN and CART have been overcome, and thus kernel methods have become more and more popular in the research fields of chemistry and biotechnology.

3.2   Kernel methods: Three key ingredients

Briefly speaking, kernel methods can be performed in two successive steps. The first step is to embed the training data in the input space into a much higher-dimensional feature space. The second step is that a linear algorithm is designed to discover the linear relationship in that feature space (see Figure 3.1). In the process, three key ingredients of kernel methods, namely the dual form of linear algorithms, nonlinear mapping, and the kernel function, should be highlighted.

3.2.1   Primal and dual forms

Many linear models for regression and classification can be reformulated in terms of a dual representation where the kernel function arises naturally. This concept plays a very important role when we consider kernel

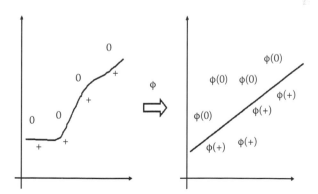

Figure 3.1 The mapping ϕ embeds the data points into a feature space where the nonlinear relationship now appears linear.

methods. To understand the dual form, let us consider a simple linear regression example: ridge regression (RR).

In simple cases, the function f can have a linear form, and then the relation between the response vector $\mathbf{y}$ and the calibration matrix $\mathbf{X}$ can be represented as follows:

$$y = f(X) + e = X \times b + e \qquad (3.1)$$

where $\mathbf{b}$ is the regression coefficient vector and $\mathbf{e}$ is the residuals. To estimate $\mathbf{b}$, RR essentially corresponds to solving the following optimization problem given by:

$$\text{minimize}: L(\mathbf{b}) = \sum_{i=1}^{n} (y_i - f(\mathbf{x}_i))^2 + \lambda ||\mathbf{b}||^2$$

$$= \sum_{i=1}^{n} (y - Xb)^t (y - Xb) + \lambda ||b||^2 \qquad (3.2)$$

where $||\mathbf{b}||^2$ denotes the norm of $\mathbf{b}$, λ denotes the positive regularization parameter that determines the trade-off between minimizing the residual sum of squares and minimizing $||\mathbf{b}||^2$. The optimal $\mathbf{b}$ can be sought by taking the derivatives of Equation (3.2) with respect to $\mathbf{b}$ and setting them equal to the zero vectors. Thus we can obtain the following equation:

$$\frac{\partial L(\mathbf{b})}{\partial \mathbf{b}} = 2X^t Xb - 2X^t y + 2\lambda b = 0 \qquad (3.3)$$

Thus, the solution of RR can be further expressed as

$$\hat{\mathbf{b}} = (X^t X + \lambda I_p)^{-1} X^t y \qquad (3.4)$$

Equation (3.4) is the commonly known solution of RR and thus is taken as the primal form of the solution. Correspondingly, the prediction function involved is given by:

$$f(\mathbf{x}) = <\hat{\mathbf{b}}, \mathbf{x}> = \mathbf{y}^t \mathbf{X}(\mathbf{X}^t \mathbf{X} + \lambda \mathbf{I}_p)^{-1} \mathbf{x} \qquad (3.5)$$

where $< \cdot >$ denotes the inner product.

Alternatively, we can also rewrite Equation (3.4) in terms of **b** to obtain

$$\hat{\mathbf{b}} = \lambda^{-1}\mathbf{X}^t(\mathbf{y} - \mathbf{X}\hat{\mathbf{b}}) = \mathbf{X}^t\boldsymbol{\alpha}$$

(3.6)

with $\boldsymbol{\alpha} = \lambda^{-1}(y - X\hat{b})$. Note that Equation (3.6) shows that the estimated regression coefficient $\hat{\mathbf{b}}$ can be written as a linear combination of the training data, say $\hat{\mathbf{b}} = \sum_{i=1}^{n}\alpha_i x_i$. If we substitute $\mathbf{X}^t\boldsymbol{\alpha}$ for $\hat{\mathbf{b}}$ in the equation $\boldsymbol{\alpha} = \lambda^{-1}\mathbf{X}^t(\mathbf{y} - \mathbf{X}\hat{\mathbf{b}})$ and then simplify Equation (3.6), $\boldsymbol{\alpha}$ can be further expressed as

$$\boldsymbol{\alpha} = (\mathbf{G} + \lambda\mathbf{I}_N)^{-1}\mathbf{y}$$

(3.7)

where $\mathbf{G} = \mathbf{X}\mathbf{X}^t$ ($n \times n$) or componentwise, $\mathbf{G}_{ij} = <x_i, x_j>$ represents the so-called *Gram matrix*, defined as an inner product of the coordinates of the training data in the *p*-dimensional space. The Gram matrix is also referred to as the kernel matrix, which is introduced in detail soon. We substitute Equation (3.7) for $\boldsymbol{\alpha}$ in (3.6) and then we can obtain

$$\hat{\mathbf{b}} = \mathbf{X}^t(\mathbf{G} + \lambda\mathbf{I}_n)^{-1}\mathbf{y}$$

(3.8)

It is surprising that $\hat{\mathbf{b}}$ has another expression form differing from Equation (3.4), a linear combination of the training data with weights $\boldsymbol{\alpha}$ (see Equation 3.8). This form is known as the dual solution of RR. The parameter $\boldsymbol{\alpha}$ is known as the dual variable. Correspondingly, the prediction function involved is given by

$$f(\mathbf{x}) = <\hat{\mathbf{b}}, \mathbf{x}> = <\sum_{i=1}^{n}\alpha_i x_i, \mathbf{x}> = \mathbf{y}^t(\mathbf{G} + \lambda\mathbf{I}_n)^{-1}\mathbf{k}$$

(3.9)

where $\mathbf{k} = <x_i, x>$ is the inner product vector between a new test sample **x** and every training sample x_i. So far, we have obtained two distinct ways to solve the RR problem: namely the primal form and the dual form. The primal form directly computes the regression coefficients **b**. This is the commonly known solution of RR. However, RR can also be solved in a dual form that only requires inner products between data points, denoted by **G** and **k**. That is, all the information RR needs in the dual form is only the inner products $<x_i, x_j>$ between data points. In the following sections, we show clearly that the inner products between data points actually construct the so-called kernel matrix, another key component of kernel methods. Thus, the use of the dual form directly gives many linear methods the outstanding advantages of the kernel function. This is why the dual form can be considered a key component of kernel methods.

3.2.2   Nonlinear mapping

It is known that RR can only model linear relationships between the calibration matrix **X** and the response vector **y**. However, in practice, their relations are often nonlinear; that is, **y** can be accurately estimated as a nonlinear function of **X**. This raises the following question, "Can we still cope with the nonlinear relation with RR?" "If we can, how should we do it?" An effective and feasible strategy is just the nonlinear mapping we want to highlight. The use of nonlinear mapping is to embed **X** into a new feature space where the linear relation can be discovered and hence RR will be able to cope with them.

Given a training set $D = \{(\mathbf{x}_1, y_1),\ (\mathbf{x}_2, y_2),\ \dots, (\mathbf{x}_n, y_n)\}$ we consider a nonlinear mapping:

$$\phi : R^p \rightarrow F \subseteq R^T$$

$$\mathbf{x} \mapsto \varphi(\mathbf{x})$$

where ϕ aims to convert nonlinear relations into linear relations. So the data points in the training set are mapped into a potentially much higher-dimensional feature space F. The training set in the feature space F can be expressed as

$$\bar{D} = \{(\varphi(\mathbf{x}_1), y_1), (\varphi(\mathbf{x}_2), y_2), \dots, (\varphi(\mathbf{x}_n), y_n)\}$$

To intuitively understand the nonlinear mapping, let us take into account a simple example. In this case, each sample is denoted by $\mathbf{x}_i = [x_{i1}, x_{i2}]$. All samples are directly depicted in 2-D input space using an ellipse $x_{i1}^2 + 2x_{i2}^2 = 4$. Figure 3.2A shows such a situation where the relation between the data points is nonlinear. Now let us consider the following mapping:

$$\phi : \mathbf{x} = [x_{i1}, x_{i2}] \in R^2 \mapsto \varphi(\mathbf{x}) = [x_{i1}^2, x_{i2}^2, \sqrt{2}x_{i1}x_{i2}] \in R^3$$

The mapping ϕ transforms the 2-D data points into a new 3-D space. While describing the data points in 3-D space, we obtain a linear relation with just two features x_{i1}^2 and x_{i2}^2 (when three features are all used, a plane will be obtained. For simplicity, we only use two features to illustrate the action of nonlinear mapping). Figure 3.2B shows the data points in 3-D

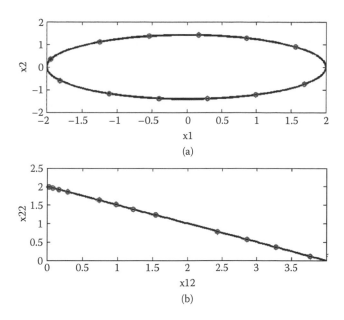

Figure 3.2 A nonlinear mapping example. We actually have an ellipse, that is, a nonlinear shape in the input space (A), and we obtain a straight line in the feature space by means of a certain nonlinear mapping.

space. As can be seen from Figure 3.2B, it is surprising that an ellipse in the input space (i.e., a nonlinear shape in the input space) is neatly transformed into a straight line in the feature space by means of nonlinear mapping ϕ.

With the help of the aforementioned example as a decision boundary, we consider a nonlinear case. In this case, the data points are class-labeled with two classes: +1 (circle) and −1 (plus). The coordinates of all data points are listed in the first two columns of Table 3.1. Figure 3.3A shows the data points in 2-D input space. From Figure 3.3A, we can clearly see that the data points are linearly inseparable except by means of a nonlinear decision function (e.g., an ellipse). To correctly classify the data points, one strategy is to embed them into a new feature space where a linear function can be sought. Herein we continue to use the above nonlinear mapping ϕ. The last three columns of Table 3.1 list the coordinates in the new feature space. All data points in the new feature space are plotted in Figure 3.3B. From Figure 3.3B, it can be seen that the data points, which are nonlinear in the original 2-D input space, remarkably have become linearly separable in 3-D feature space. Accordingly, the quadratic decision boundary in 2-D input space is transformed into a linear decision boundary. That is, the nonlinear mapping takes the data points from a 2-D input space to a 3-D feature space in a way that the linear relation

Table 3.1 A Simulated Example Used for Explaining Nonlinear Mapping

Label	$x_1 \times 10^3$	$x_2 \times 10^3$	$x_1^2 \times 10^3$	$x_2^2 \times 10^3$	$\sqrt{2}\,x_1 x_2 \times 10^3$
1	73.59	14.49	5.415	0.2098	1.508
1	15.58	54.01	0.2426	2.918	1.190
1	−12.17	−18.96	0.1480	0.3594	0.3262
1	60.98	−58.49	3.718	3.421	−5.043
1	5.490	−91.93	0.03014	8.452	−0.714
1	−72.70	−18.96	5.285	0.3594	1.949
1	−39.91	63.14	1.593	3.986	−3.564
−1	−113.1	163.5	12.78	26.72	−26.14
−1	−161.0	117.9	25.91	13.89	−26.83
−1	−186.2	8.406	34.67	0.07066	−2.214
−1	−186.2	−94.97	34.67	9.020	25.01
−1	−87.83	−210.5	7.714	44.31	26.15
−1	55.93	−225.7	3.129	50.95	−17.86
−1	192.1	−143.6	36.91	20.63	−39.02
−1	242.6	41.85	58.84	1.752	14.36
−1	197.1	175.6	38.88	30.85	48.98
−1	15.58	224.3	0.2426	50.30	4.941

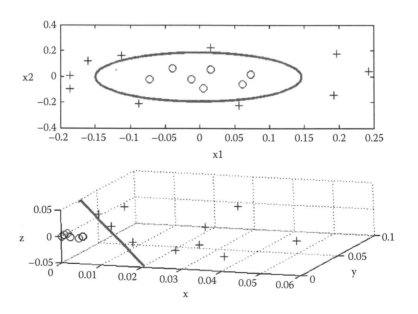

Figure 3.3 The linearly inseparable data points in the original input space have been linearly separable in the feature space by means of nonlinear mapping.

in the feature space corresponds to the quadratic relation in the input space.

As mentioned above, changing the data representation via nonlinear mapping can significantly simplify data analysis. Generally speaking, we can always seek the linear representation of data by means of specific nonlinear mapping. Instead of the original input space, we can consider the same algorithms in the new feature space F. If $\phi : R^p \rightarrow F$ is a nonlinear mapping, we can learn nonlinear relations with linear methods in the feature space.

$$f(\mathbf{x}) = \sum_{i=1}^{T} b_i \varphi_i(\mathbf{x}) + b_0 \qquad (3.10)$$

where T is the number of the dimensionality of the feature space. In other words, the nonlinear methods can be constructed in two successive steps: first a nonlinear mapping transforms the data into a feature space F, and then a linear learning algorithm is designed to discover the linear data relations in that space. In the process, the nonlinear mapping aims to convert nonlinear relations to linear relations and hence reflects our expectations about the relation to be learned.

Although the primal form of the solution could be used in the feature space, the computational complexity of Equation (3.5) will be very high if T is very large. It requires solving the large $T \times T$ system for the inverse of $\varphi(\mathbf{X})^t \varphi(\mathbf{X})$. However, in the dual representation, the prediction function $f(\mathbf{x}) = \mathbf{y}^t (\mathbf{G} + \lambda \mathbf{I}_n)^{-1} \mathbf{k}$ of RR involves the Gram matrix $\mathbf{G}$ with entries $\mathbf{G}_{ij} = <\varphi(\mathbf{x}_i), \varphi(\mathbf{x}_j)>$ after a nonlinear mapping. That is, all the information RR needs is the inner products between data points $<\varphi(\mathbf{x}_i), \varphi(\mathbf{x}_j)>$ in the feature space F. To obtain the final prediction function, we only need to compute the Gram matrix $\mathbf{G}$ and inner product vector $\mathbf{k}$ in the dual representation. Thus the dual solution in the feature space only requires solving the $n \times n$ system rather than the $T \times T$ system. This is very effective when T is far larger than n, a situation quite common in chemometrics. The use of the dual form can greatly improve the computational efficiency of the kernel methods. This is also the other reason why we use the dual form in the kernel methods.

3.2.3   Kernel function and kernel matrix

In the above analysis, no matter whether it is the primal form or the dual form, we require an implicit mapping of the data to the feature space F when dealing with nonlinear relations with linear methods. Generally speaking, such implicit mapping actually reflects our expectations about the relation to be learned and therefore is usually very difficult to obtain. Furthermore, the adjustment of the parameters involved with nonlinear mapping may be a time-consuming and tedious task. There is, however, a simple way of calculating the inner product in the feature space directly as a function of the original input variables. The direct computation approach is called the kernel function approach. More formally, the kernel function is defined in such a way, that for all x_i, $x_j \in X$,

$$\kappa(x_i, x_j) = <\varphi(x_i), \varphi(x_j)> \tag{3.11}$$

where ϕ is a mapping from X to the feature space F. To understand how the kernel function works, let us continue to consider the mapping $\phi : x = [x_{i1}, x_{i2}] \in R^2 \mapsto \varphi(x) = [x_{i1}^2, x_{i2}^2, \sqrt{2}x_{i1}x_{i2}]$. The composition of the feature map with inner product in the feature space F can be evaluated as follows:

$$< \varphi(x_i), \varphi(x_j) > \: = \: < [x_{i1}^2, x_{i2}^2, \sqrt{2}x_{i1}x_{i2}], \, [x_{j1}^2, x_{j2}^2, \sqrt{2}x_{j1}x_{j2}]$$

$$= x_{i1}^2 x_{j1}^2 + x_{i2}^2 x_{j2}^2 + 2x_{i1}x_{i2}x_{j1}x_{j2}$$

$$= <[x_{i1}, x_{i2}], [x_{j1}, x_{j2}]>^2$$

$$= <x_i, x_j>^2$$

Hence, the function $\kappa(x_i, x_j) = <x_i, x_j>^2$ is a kernel function with its corresponding feature space F. Thus we can directly compute the inner product $<\varphi(x_i), \varphi(x_j)>$ between the projections of two data points into the feature space without explicitly evaluating their coordinates $\varphi(x_i)$ and $\varphi(x_j)$. The inner product in the feature space is directly calculated as the quadratic function $<x_i, x_j>^2$ of the inner product in the input space. In this process, we even do not need to know the underlying feature mapping $\phi : x = [x_{i1}, x_{i2}] \in R^2 \mapsto \varphi(x) = [x_{i1}^2, x_{i2}^2, \sqrt{2}x_{i1}x_{i2}] \in R^3$. It is worth noting that the kernel function is an inner product between two points in the new feature space. Moreover, it is also a quite simple function of the data points in the original input space. This means that the inner products of the mapped data points in the feature space can be calculated with some functions of the data points in the input space by means of kernel functions.

The use of the kernel function is an attractive computational shortcut. A curious fact about using a kernel is that we do not need to know the underlying feature mapping that can learn in the feature space. In practice, the approach adopted is to define a kernel function directly, hence implicitly to define the feature space. In this way, we avoid the feature space not only in the computation of the inner product, but also in the design of the learning algorithm itself. So kernels allow us to compute inner products in the feature space, where one could otherwise not be able to perform any computations. The use of the kernel function discovers a very effective and feasible way to combine the nonlinear mapping with the dual form of the linear algorithms in nonlinear chemical systems.

As can be seen above, the kernel function plays a key role in the construction of the kernel methods. Then, it seems to be necessary for us to know what properties of a kernel function $K(x, z)$ ensure that it is a kernel for some feature space. Clearly, the function must first be symmetric,

$$K(x, z) = \langle \varphi(x), \varphi(z) \rangle = \langle \varphi(z), \varphi(x) \rangle = K(z, x)$$

and satisfy the inequality that follows from the Cauchy–Schwarz inequality.

$$K(x, z)^2 = \langle \varphi(x), \varphi(z) \rangle^2 \leq ||\varphi(x)||^2 ||\varphi(z)||^2$$
$$= \langle \varphi(x), \varphi(x) \rangle \langle \varphi(z), \varphi(z) \rangle = K(x, x)K(z, z)$$

But the conditions are not sufficient to guarantee the existence of a feature space. Then we would like to know what additional properties are also required to warrant sufficient conditions. The theoretical derivation can prove that if a function satisfies the Mercer condition, it can be regarded as a kernel function [1].

Theorem 3.1 (Mercer) Let X be a compact subset of R^p. Suppose K is a continuous symmetric function such that the integral operator T_K: $L_2(X) \rightarrow L_2(X)$.

$$(T_K f)(\bullet) = \int_X K(\bullet, x) f(x) dx$$

is positive; that is,

$$\int_{X \times X} K(x, z) f(x) f(z) dx dz \geq 0$$

for all $f \in L_2(\mathbf{X})$. Then we can expand $\mathbf{K}(\mathbf{x}, \mathbf{z})$ in a uniformly convergent series (on $\mathbf{X} \times \mathbf{X}$) in terms of T_Ks eigenfunctions $\varphi_j \in L_2(\mathbf{X})$, normalized in such a way that $\|\varphi_j\|_{L_2} = 1$, and positive associated eigenvalues $\lambda_j \geq 0$.

$$\mathbf{K}(\mathbf{x}, \mathbf{z}) = \sum_{j=1}^{\infty} \lambda_j \varphi_j(\mathbf{x}) \varphi_j(\mathbf{z})$$

Let us observe that this theorem, the positivity condition $\int_{\mathbf{X} \times \mathbf{X}} \mathbf{K}(\mathbf{x}, \mathbf{z}) f(\mathbf{x}) f(\mathbf{z}) \, d\mathbf{x} d\mathbf{z} \geq 0$, $\forall f \in L_2(\mathbf{X})$, corresponds to the positive semidefinite condition in the finite case. This gives the second characterization of a kernel function that is proved to be most useful when we construct kernels. We use the term kernel to refer to the function satisfying this property, but in the literature these are often called Mercer kernels.

Based on the condition stated in the Mercer theorem, a number of other properties of the kernel can be obtained. Assume that $\mathbf{K}_1$, $\mathbf{K}_2$, and $\mathbf{K}_3$ are kernels; B is a symmetric positive-semidefinite matrix; and $p(\mathbf{x})$ is a polynomial with positive coefficients. Then the following functions are also kernels.

1. $\mathbf{K}(\mathbf{x}, \mathbf{z}) = \mathbf{K}_1(\mathbf{x}, \mathbf{z}) + \mathbf{K}_2(\mathbf{x}, \mathbf{z})$

2. $\mathbf{K}(\mathbf{x}, \mathbf{z}) = a\mathbf{K}_1(\mathbf{x}, \mathbf{z})$

3. $\mathbf{K}(\mathbf{x}, \mathbf{z}) = \mathbf{K}_1(\mathbf{x}, \mathbf{z})\mathbf{K}_2(\mathbf{x}, \mathbf{z})$

4. $\mathbf{K}(\mathbf{x}, \mathbf{z}) = f(\mathbf{x})f(\mathbf{z})$

5. $\mathbf{K}(\mathbf{x}, \mathbf{z}) = \mathbf{K}_3 \langle \varphi(\mathbf{x}), \varphi(\mathbf{z}) \rangle$

6. $\mathbf{K}(\mathbf{x}, \mathbf{z}) = \mathbf{x}^t \mathbf{B} \mathbf{z}$

7. $\mathbf{K}(\mathbf{x}, \mathbf{z}) = p(\mathbf{K}_1(\mathbf{x}, \mathbf{z}))$

8. $\mathbf{K}(\mathbf{x}, \mathbf{z}) = \exp(\mathbf{K}_1(\mathbf{x}, \mathbf{z}))$

9. $\mathbf{K}(\mathbf{x}, \mathbf{z}) = \exp(-\|\mathbf{x} - \mathbf{z}\|^2 / \delta^2)$

Thus, instead of the above-mentioned linear kernel function, other kernel functions can be implemented to cope with the data at hand. Table 3.2 lists four commonly used kernel functions in chemometrics. If the chemical data are linear or nearly linear, we can first select the linear kernel function because it usually exhibits good prediction performance. For the nonlinear chemical data, the radial basis kernel function should be tried in computation at first. With the aid of these kernel functions, we

Table 3.2 Commonly Used Kernel Functions

Kernel	$K(x_1, x_2)$
Linear kernel	$x_1^t x_2$
Polynomial kernel	$[scale \cdot (x_1^t x_2) + offset]^{degree}$
Radial basis kernel	$exp(-\delta \lVert x_1 - x_2 \rVert^2)$
Sigmoidal kernel	$tanh[scale \cdot (x_1^t x_2) + offset]$

Note: Scale, offset, degree, and δ are the parameters of used kernels, which should be predefined by the user. Among different types of kernels, the radial basis kernel is more common.

reconsider the solution of RR (see Equation (3.9)). When we replace **G** by kernel matrix **K**, denoted in Table 3.2, a kernel RR can be naturally established to tackle nonlinear chemical problems.

3.3 Modularity of kernel methods

So far, we have highlighted these key components of kernel methods in detail. The use of the dual form makes linear algorithms able to be solved only by inner product information (i.e., the kernel function). The use of nonlinear mapping makes kernel methods able to treat nonlinear problems with linear algorithms. The kernel function is also able to effectively bridge the gaps between the dual form and nonlinear mapping. Thus we have clearly seen that kernel RR can naturally be established. A direct consequence of this finding is that every (linear) algorithm that only uses inner products can implicitly be executed in *F* by using kernels; that is, one could elegantly construct a nonlinear version of a linear algorithm [22]. This is the so-called modularity of kernel methods that we introduce, which indicates that all kernel-based methods have a consistent framework.

Clearly, we can use any suitable kernel to represent the data being considered. Similarly, we can use any linear algorithm, which only uses inner products between samples, to establish a model while retaining the chosen kernel. This illustrates that we can consider the modeling methods separately from the choice of the kernel functions. Figure 3.4 shows the steps involved in the implementation of the kernel methods. The training samples are first represented using a kernel function to create a kernel matrix, and subsequently the kernel matrix obtained is modeled by a linear algorithm to produce a complex model. The model is finally used to predict the unseen samples. In the process, the linear algorithms are naturally combined with the specific kernel function to produce a more complex algorithm in a high-dimensional feature space.

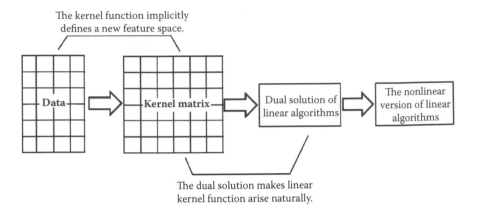

Figure 3.4 The general steps used to implement kernel methods.

In the following sections we use this philosophy to show how one could deal with nonlinear problems by some commonly used linear modeling algorithms. We first reformulate linear, inner product-based algorithms in the input space (i.e., the dual form in the input space). The dual representation in the input space makes the linear kernel matrix arise naturally. Second, the original data are mapped into an implicit feature space. Correspondingly, the linear kernel matrix can be transformed to the nonlinear kernel matrix by means of the specific kernel function. Finally one is able to generate powerful nonlinear algorithms.

3.4   *Kernel principal component analysis*

Primal principal component analysis (PCA) is a linear algorithm [23,24]. It can extract linear structures in the datasets. For p-dimensional variable data, a set of orthogonal directions, usually less than p, capturing most of the variance in the data can be obtained. In practice, one typically wants to describe the data with reduced dimensionality by extracting a few meaningful components, while at the same time one also wants to retain most existing structure in the data. That is, the first k projections are usually used to reconstruct the data with minimal squared errors. The technique, say the eigenvalue and eigenvector problem, is usually used to fulfill this task (see Table 3.3). Unfortunately, for those data with nonlinear structures, the primal PCA may not be competent for the purpose. Thus, kernel PCA (KPCA) is the natural version of PCA in a kernel feature space for dealing with the nonlinearity [25–28]. The basic idea of KPCA is to map the original dataset into some higher-dimensional feature space where we use PCA to establish a linear relationship which, however, is nonlinear in the original input space (see Table 3.3).

Table 3.3 Comparison of Primal PCA and KPCA

	PCA	KPCA
1	$\mathbf{X} = \mathbf{X} - \text{ones}(n,1) \times \text{mean}(\mathbf{X})$	$\mathbf{K} = \kappa(\mathbf{x},\mathbf{z}) = \varphi(\mathbf{X})\varphi(\mathbf{X})^t$
2	$\mathbf{C} = \mathbf{X}^t\mathbf{X}/n$	$\mathbf{K} = (1/n)\mathbf{jj}^t\mathbf{K} - (1/n)\mathbf{Kjj}^t + (1/n^2)(\mathbf{j}^t\mathbf{Kj})\mathbf{jj}^t$
3	$[\mathbf{V}, \Lambda] = \text{eig}(\mathbf{C})$	$[\mathbf{U}, \Lambda] = \text{eig}(\mathbf{K})$
4		$\mathbf{V}_k = \mathbf{X}^t\mathbf{U}_k\Lambda_k^{-1/2} \ (\alpha = \mathbf{U}\Lambda^{-1/2})$
5	$\mathbf{X}_{\text{new}} = \mathbf{X}_{\text{new}}\mathbf{V}_k$	$\mathbf{X}_{\text{new}} = \mathbf{X}_{\text{new}}\mathbf{V}_k \ (\mathbf{x}_{\text{new}} = \kappa(\mathbf{x}_i, \mathbf{x}_{\text{new}})\alpha_k)$

j represents a vector with all elements equal to 1.

For primal PCA, one can first compute the covariance matrix $\mathbf{C}$:

$$\mathbf{C} = \frac{1}{n}\sum_{i=1}^{n} \mathbf{x}_i\mathbf{x}_i^t = \frac{1}{n}\mathbf{X}^t\mathbf{X} \tag{3.12}$$

A principal component $\mathbf{v}$ is then computed by solving the following eigenvalue problem:

$$\lambda\mathbf{v} = \mathbf{C}\mathbf{v} = \frac{1}{n}\mathbf{X}^t\mathbf{X}\mathbf{v} \tag{3.13}$$

Here $\lambda > 0$ and $\mathbf{v} \neq 0$. All eigenvectors with nonzero eigenvalues must be in the span of the data (i.e., dual form). Thus, the eigenvectors can be written as

$$\mathbf{v} = \sum_{i=1}^{n} \alpha_i\mathbf{x}_i = \mathbf{X}^t\alpha \tag{3.14}$$

where $\alpha = [\alpha_1, ..., \alpha_n]^t$, by substituting $\mathbf{v}$ in Equation (3.13) for Equation (3.14), that is,

$$\lambda\mathbf{X}^t\alpha = \frac{1}{n}\mathbf{X}^t\mathbf{X}\mathbf{X}^t\alpha$$

Then the eigenvalue problem can be represented by the simple form:

$$\lambda\alpha = \frac{1}{n}\mathbf{K}\alpha \tag{3.15}$$

where $\mathbf{K}_{ij} = \mathbf{x}_i^t\mathbf{x}_j$ or $\mathbf{K} = \mathbf{X}\mathbf{X}^t \in R^{n \times n}$ is a linear kernel matrix. To ensure the normality of the principal component (i.e., $||\mathbf{v}||^2 = 1$) the calculated α must be scaled such to satisfy $||\alpha||^2 = 1/(n\lambda)$.

To derive KPCA, one first needs to map the data $\mathbf{X} = [\mathbf{x}_1, \mathbf{x}_2, ..., \mathbf{x}_n]^t \in R^p$ into a feature space F (i.e., $\mathbf{M} = [\varphi(\mathbf{x}_1), \varphi(\mathbf{x}_2), ... , \varphi(\mathbf{x}_n)]^t$). Hence, we can easily obtain $\mathbf{K} = \mathbf{M}\mathbf{M}^t \in R^{n \times n}$. Such a nonlinear kernel matrix $\mathbf{K}$ can be directly generated by means of a specific kernel function (denoted by Table 3.2).

For extracting features of a new sample $\mathbf{x}$ with KPCA, one simply projects the mapped sample $\varphi(\mathbf{x})$ onto the first k projections $\mathbf{V}_k$,

$$\mathbf{V}_k \cdot \varphi(\mathbf{x}) = \sum_{i=1}^{n} \alpha_i^k < \varphi(\mathbf{x}_i), \varphi(\mathbf{x}) > = \sum_{i=1}^{n} \alpha_i^k \kappa(\mathbf{x}_i, \mathbf{x}) = \mathbf{k}_{test} \cdot \alpha^k \qquad (3.16)$$

where $\mathbf{k}_{test} = \mathbf{M}\varphi(\mathbf{x}) = [\kappa(\mathbf{x}_1, \mathbf{x}), \kappa(\mathbf{x}_2, \mathbf{x}),..., \kappa(\mathbf{x}_n, \mathbf{x})]$.

It seems that we need here a simple example to show what the primal PCA and KPCA mean and how they work. Figure 3.5 shows a nonlinear example in which a sphere with radius equal to 1, denoted by the black circle, is regarded as one class and randomly generated points ranging from −0.5 to 0.5, denoted by the gray circle, are regarded as the other class. To begin with, the primal PCA is employed to visualize the trends of the data in two-dimensional space, as described in the left panel of Figure 3.6. Clearly, no separation can be observed in the first two scores. It seems

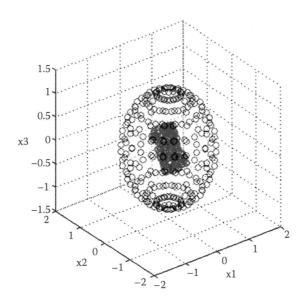

Figure 3.5 A linearly inseparable example used in PCA and KPCA. The data points denoted by a black circle are generated using a sphere with radius equal to 1, and the data points denoted by a gray circle are generated using random number from −0.5 to 0.5.

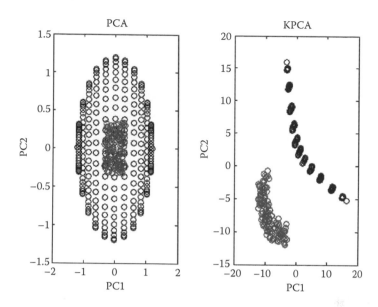

Figure 3.6 The score plot of PCA and KPCA. The primal PCA is not capable of distinguishing the two classes (the left plot). However, a big separation between two classes can be obtained using KPCA (the right plot). The use of a Gaussian kernel function makes the nonlinear data linearly separable in KPCA.

that the primal PCA is not capable of differentiating the two classes with specific nonlinear structure because inherently it is a linear algorithm. The right panel of Figure 3.6 shows the results of applying KPCA to the data. KPCA maps the original data with nonlinear features in the input space into the kernel feature space in which the linear PCA algorithm is then performed. Here, the first two principal components in KPCA are extracted with the Gaussian kernel function. As can be seen from Figure 3.6, the first two principal components are able to distinguish the two classes nicely. From the example, one can obtain that KPCA is more suitable to describe the nonlinear structure than PCA.

3.5   Kernel partial least squares

Kernel partial least squares (KPLS) [29] is a nonlinear extension of linear PLS in which the training samples are transformed into a feature space via a nonlinear mapping. The primal PLS algorithm can then be carried out in the feature space. In chemometrics, the nonlinear iterative partial least squares (NIPALS) algorithm is the commonly used PLS version [30,31]. However, the modified NIPALS algorithm [32], proposed by Lewi, normalizes the latent vectors $\mathbf{t}$, $\mathbf{u}$ rather than the weight vectors $\mathbf{w}$, $\mathbf{c}$. The modified PLS algorithm is shown in the left column of Table 3.4. By the

Table 3.4 Comparison of Primal PLS and KPLS

	PLS	KPLS
1	Randomly initiate $\mathbf{u}$ or $\mathbf{u} = \mathbf{y}$	Randomly initiate $\mathbf{u}$ or $\mathbf{u} = \mathbf{y}$
2	$\mathbf{W} = \mathbf{X}^t\mathbf{u}$	
3	$\mathbf{t} = \mathbf{Xw}$	$\mathbf{t} = \varphi(\mathbf{X})\varphi(\mathbf{X})^t\,\mathbf{u} = \mathbf{Ku}$
	$\mathbf{t} = \mathbf{t}/\|\mathbf{t}\|$	$\mathbf{t} = \mathbf{t}/\|\mathbf{t}\|$
4	$\mathbf{c} = \mathbf{y}^t\mathbf{t}$	$\mathbf{c} = \mathbf{y}^t\mathbf{t}$
5	$\mathbf{u} = \mathbf{yc}$	$\mathbf{u} = \mathbf{yc}$
	$\mathbf{u} = \mathbf{u}/\|\mathbf{u}\|$	$\mathbf{u} = \mathbf{u}/\|\mathbf{u}\|$
6	Repeat 2–5 until convergence	Repeat 2–5 until convergence
7	$\mathbf{X} = \mathbf{X} - \mathbf{tt}^t\mathbf{X}$	$\mathbf{K} = (\mathbf{I} - \mathbf{tt}^t)\mathbf{K}(\mathbf{I} - \mathbf{tt}^t)$
	$\mathbf{y} = \mathbf{y} - \mathbf{tt}^t\mathbf{y}$	$\mathbf{y} = \mathbf{y} - \mathbf{tt}^t\mathbf{y}$

connection of Steps 2 and 3 and by using the $\varphi(\mathbf{X})$ matrix of the mapping samples, we can derive the algorithm for the nonlinear kernel PLS model, which is listed in the right column of Table 3.4.

Applying the so-called "kernel trick," that is, Equation (3.11), we can clearly see that $\mathbf{MM}^t$ represents the kernel matrix of the inner products between all mapped input samples. Thus, instead of an explicit nonlinear mapping, the kernel function can be used to modify line 7 in Table 3.4. Instead of the deflation of the $\mathbf{X}$ matrix, the deflation of kernel matrix is given by:

$$K \leftarrow (I - tt^t)K(I - tt^t) = K - tt^t\,K - K\,tt^t + tt^t\,K\,tt^t \qquad (3.17)$$

where $\mathbf{I}$ is an n-dimensional identity matrix. The estimated regression coefficient $\mathbf{b}$ in the KPLS algorithm can be given by:

$$b = M^t\,U(T^tKU)^{-1}T^ty \qquad (3.18)$$

where $\mathbf{U}$ and $\mathbf{T}$ consist of $\mathbf{u}$ and $\mathbf{t}$ in the column form, respectively. For new test data consisting of n_t samples, one can use the following equations to predict the training set and test data, respectively:

$$\hat{\mathbf{y}} = \mathbf{Mb} = \mathbf{KU}(\mathbf{T}^t\mathbf{KU})^{-1}\mathbf{T}^t\mathbf{y} \qquad (3.19)$$

$$\hat{\mathbf{y}}_t = \mathbf{M}_{test}\mathbf{b} = \mathbf{K}_t\mathbf{U}(\mathbf{T}^t\mathbf{KU})^{-1}\mathbf{T}^t\mathbf{y} \qquad (3.20)$$

where $\mathbf{M}_{test}$ is the matrix of the mapped test points and $\mathbf{K}_{test}$ is the ($n_{test} \times n$) test kernel matrix whose elements are $\kappa(\mathbf{x}_i, \mathbf{x}_j)$, where $\mathbf{x}_i$ is the ith test sample and $\mathbf{x}_j$ is the jth training sample.

3.6 Kernel Fisher discriminant analysis

Fisher discriminant analysis (FDA) finds a linear projection of the data onto a one-dimensional subspace such that the classes are well separated according to a certain measure of separability (see Table 3.5) [33–35]. However, FDA seems to be very limited in that it is clearly beyond its capabilities to separate the data with nonlinear structures. Fortunately, it is possible to have both linear models and a very rich set of nonlinear decision functions by means of the kernel trick [36–39]. The idea of kernel FDA (KFDA) is to solve the problem of FDA in a kernel feature space, thereby yielding a nonlinear discriminant in the input space (see Table 3.5).

For primal FDA, it can be achieved by maximizing the following Fisher criterion:

$$J(w) = \frac{\mathbf{w}^t \mathbf{S}_B \mathbf{w}}{\mathbf{w}^t \mathbf{S}_W \mathbf{w}}$$

(3.21)

where $\mathbf{S}_B$ and $\mathbf{S}_W$ are the between-class and the total scatter matrices defined in the original input space, respectively. They have the following expression:

$$\mathbf{S}_B = (\mathbf{m}_2 - \mathbf{m}_1)(\mathbf{m}_2 - \mathbf{m}_1)^t \text{ and } \mathbf{S}_W = \sum_{k=1}^{2} \sum_{i=1}^{n_k} (\mathbf{x}_i - \mathbf{m}_k)(\mathbf{x}_i - \mathbf{m}_k)^t$$

Here $\mathbf{m}_k$ and n_k denote the sample mean value and the number of samples for class k, respectively. According to the dual form, the optimal discriminant vector $\mathbf{w}$ can be expressed as a linear combination of the training samples in the original input space.

$$\mathbf{w} = \sum_{i=1}^{n} \alpha_i \mathbf{x}_i = \mathbf{X}^t \boldsymbol{\alpha}$$

(3.22)

Table 3.5 Comparison of Primal FDA and KFDA

	FDA	KFDA
1	$\mathbf{X} = \mathbf{X} - \text{ones}(n,1) \times \text{mean}(\mathbf{X})$	$\mathbf{K} = \kappa(\mathbf{x},\mathbf{z}) = \mathbf{XX}^t \text{ or } \varphi(\mathbf{X})\varphi(\mathbf{X})^t$
2		$\mathbf{K} = (1/n)\mathbf{jj}^t\mathbf{K} - (1/n)\mathbf{Kjj}^t + (1/n^2)(\mathbf{j}^t\mathbf{Kj})\mathbf{jj}^t$
3	Compute $\mathbf{S}_B$ and $\mathbf{S}_W$	Compute $\mathbf{KWK}$ and $\mathbf{KK}$
4	$[\mathbf{V},\Lambda] = \text{eig}(\mathbf{S}_B, \mathbf{S}_W)$	$[\mathbf{V},\Lambda] = \text{eig}(\mathbf{KWK}, \mathbf{KK})$
5	$y = \mathbf{X}_{\text{new}}\mathbf{V}_1$	$y = \mathbf{k}_{\text{test}}\, \boldsymbol{\alpha}\,(\boldsymbol{\alpha} = \mathbf{V}_1, \mathbf{k}_{\text{test}} = \mathbf{M}\varphi(\mathbf{x}_{\text{test}}))$

$\mathbf{j}$ represents a vector with all elements equal to 1.

where $\mathbf{X} = [\mathbf{x}_1, \mathbf{x}_2, \ldots, \mathbf{x}_n]^t$ and $\boldsymbol{\alpha} = [\alpha_1, \ldots, \alpha_n]^t$. Substituting Equation (3.22) into (3.21), we have

$$J(\boldsymbol{\alpha}) = \frac{\boldsymbol{\alpha}^t \mathbf{KWK}\boldsymbol{\alpha}}{\boldsymbol{\alpha}^t \mathbf{KK}\boldsymbol{\alpha}} \qquad (3.23)$$

where $\mathbf{K} = \mathbf{XX}^t \in R^{n \times n}$ or $\mathbf{K}_{ij} = \mathbf{x}_i^t \mathbf{x}_j$ is a linear kernel matrix; $\mathbf{W} = \text{diag}(\mathbf{W}_1, \mathbf{W}_2)$; and $\mathbf{W}_k$ is an $n_k \times n_k$ matrix with terms all equal to $1/n_k$. To maximize Equation (3.23) one could solve the generalized eigenvalue problem $\mathbf{KWK}\boldsymbol{\alpha} = \lambda\mathbf{KK}\boldsymbol{\alpha}$ and then select eigenvector $\boldsymbol{\alpha}$ with maximal eigenvalue λ.

Now, to derive KFDA one first needs to map the data $\mathbf{X} = [\mathbf{x}_1, \mathbf{x}_2, \ldots, \mathbf{x}_n]^t \in R^p$ into the kernel feature space F; that is, $\mathbf{M} = [\varphi(\mathbf{x}_1), \varphi(\mathbf{x}_2), \ldots, \varphi(\mathbf{x}_n)]^t$. Thus, substituting $\varphi(\mathbf{x})$ for all $\mathbf{x}$ in Equation (3.21), we can similarly obtain the optimal solution for KFDA in the feature space. Here, $\mathbf{K} = \mathbf{MM}^t$. Such a nonlinear kernel matrix $\mathbf{K}$ can be generated directly by means of a specific kernel function (see Table 3.2).

Correspondingly, given a test sample $\mathbf{x}$, the projection of this sample onto the discriminant vector can be computed by:

$$\mathbf{w} \cdot \varphi(\mathbf{x}_{\text{test}}) = \sum_{i=1}^{n} \alpha_i \kappa(\mathbf{x}_i, \mathbf{x}_{\text{test}}) = \mathbf{k}_{\text{test}}\boldsymbol{\alpha} \qquad (3.24)$$

where $\mathbf{k}_{\text{test}} = \mathbf{M}\varphi(\mathbf{x}) = [\kappa(\mathbf{x}_1, \mathbf{x}), \kappa(\mathbf{x}_2, \mathbf{x}), \ldots, \kappa(\mathbf{x}_n, \mathbf{x})]$.

3.7   Relationship between kernel function and SVMs

As we discussed in the last chapter, SVMs were originally developed by Vapnik and his coworkers and they have shown promising capability for solving both linear and nonlinear problems. For linearly separable cases, SVC performs classification tasks by constructing a hyperplane in the input space to differentiate two classes with a maximum margin. Note that the significant feature of linear SVC is that it attempts to seek a "safest" hyperplane maximizing the square sum of the distance between the hyperplane and all data points. The "safest" hyperplane can give the correct prediction as soon as possible when new samples arrive. That is, linear SVC can define the "safest" hyperplane to give the best prediction performance.

In order to deal with nonlinearity embedded in the data, the kernel function was introduced into SVMs. In most cases, the data are linearly inseparable due to two reasons. One is noise contamination, unknown background, or man-made mistakes, and the other is intrinsic nonlinearity. To deal with the two situations, as shown in Chapter 2, slack variable and kernel functions were cleverly introduced by Cortes and Vapnik. To allow for specific training error, slack variables are introduced to consider the inevitable measured errors in data to a certain extent. These slack variables are associated with the misclassified samples. Even though erroneous classification cannot be avoided, the effect of the misclassified samples can be reduced by means of these slack variables. Thus, the optimal object of SVMs can further be re-expressed in the following form with a slack variable ξ introduced:

$$\text{minimize}: \frac{1}{2}||\mathbf{w}||^2 + C\sum_{i=1}^{n}(\xi_i) \tag{3.25}$$

$$\text{subject to}: \begin{array}{l} y_i(\mathbf{w}^t\mathbf{x}_i + b) \geq 1 - \xi_i \\ \xi_i \geq 0 \quad i = 1, 2, \ldots, n \end{array}$$

where C is a penalizing factor that controls the trade-off between the training error and the margin. As indicated in Section 3.1, Equation (3.25) can be rewritten as the dual form given by:

$$\text{maximize}: \sum_{i=1}^{n}\alpha_i - \frac{1}{2}\sum_{i,j=1}^{n}\alpha_i\alpha_j y_i y_j \mathbf{x}_i^t \mathbf{x}_j \tag{3.26}$$

$$\text{subject to}: \begin{array}{l} 0 \leq \alpha_i \leq C, \quad i = 1, \ldots, n \\ \sum_{i=1}^{n}\alpha_i y_i = 0 \end{array}$$

where α_i and α_j are the optimized Lagrange multipliers. Equation (3.26) can be solved by means of quadratic optimization methods. Thus, the decision function of SVC can be written as

$$f(\mathbf{x}) = \text{sign}(\mathbf{w}^t\mathbf{x} + b) = \text{sign}\left(\sum_{i=1}^{n} y_i\alpha_i < \mathbf{x}_i, \mathbf{x} > + b\right) \tag{3.27}$$

where α_i ($0 < \alpha_i \le C$) is the solution of the QP problem in Equation (3.26), and b can be computed using the following equation:

$$b = y_j - \sum_{i=1}^{n} y_i \alpha_i < x_i, x_j >, j \in \left\{ j \mid 0 < \alpha_j \le C \right\} \tag{3.28}$$

When the data structure is intrinsically nonlinear, the kernel function can be performed to extend the linear problems to nonlinear problems. Thus, when we replace the $<x_i, x>$ term in Equation (3.27) with K (x_i, x), a nonlinear version of SVC can easily be constructed. For SVR, the same tools (i.e., slack variables and kernel function) can be used. As mentioned above, SVMs can also be further extended and solved according to three key components of kernel methods. They have similar ways as kernel RR except for the loss function (i.e., Equations (3.26) and (3.2)). The use of the kernel function is one of the chief factors in SVM becoming an effective tool of machine learning.

It seems to be quite necessary to show how to cope with the nonlinear classification examples using the aforementioned three kernel methods, namely SVM, KPLS, and KFDA. A simple nonlinear classification example with 50 samples in each class is shown in Figure 3.7. As can be shown from the figure, the data points are linearly inseparable unless a nonlinear

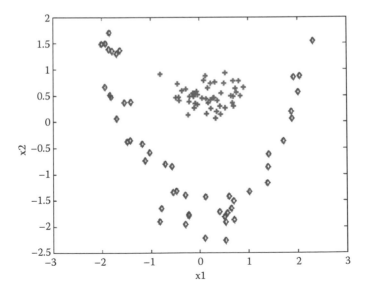

Figure 3.7 A simulated classification example used to show the good performance of kernel methods.

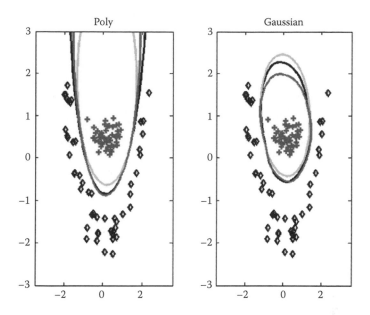

Figure 3.8 The decision boundaries of three kernel methods with polynomial and Gaussian kernel functions. SVM: light gray, KPLS: dark gray, KFDA: black.

decision function, such as a conic or an ellipse, is used. So, SVM, KPLS, and KFDA are used to establish the nonlinear models, respectively.

The results of three kernel methods with polynomial and Gaussian kernel functions are represented in Figure 3.8. It is shown that the three methods can all classify the data points correctly. However, there seem to be some slight differences among the three kernel methods. Furthermore, we can also see that two different decision boundaries are obtained using polynomial and Gaussian kernel functions. The choice of kernel functions is very important for generating a correct decision boundary because it is closely related to the underlying structure of the data. Herein, for the polynomial kernel function, the polynomial function degree is chosen as 2 for the three kernel methods. For the Gaussian kernel function, its widths are set to 1, 0.6, and 0.8 for SVM, KPLS, and KFDA, respectively.

In addition, KPLS can also project the data points in the feature space into a reduced latent variable space where we can directly observe the underlying data structure. Figure 3.9 shows the result of the score plot in KPLS. We can clearly see from Figure 3.9 that these linearly inseparable data points in the original input space have been linearly separable in the latent variable space, which adequately indicates the primal PLS model can indeed model the nonlinear relationship in the data by means of the action of kernel functions.

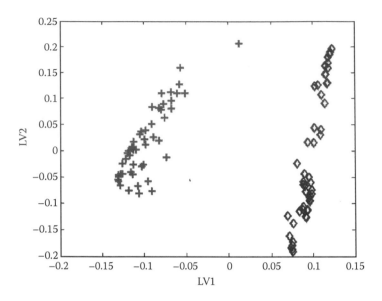

Figure 3.9 The score plot of kernel PLS.

3.8   *Kernel matrix pretreatment*

To obtain better prediction performance, mean centering of the data should be carried out in the feature space for every algorithm. Given an implicit mapping $\phi: \mathbf{X} \in R^p \rightarrow \phi(\mathbf{X}) \in F$, the mean centering of the data in the feature space can be accomplished conveniently by means of the following algorithm [22]:

$$\mathbf{H} = \mathbf{I} - \frac{1}{n}\mathbf{1}_n\mathbf{1}_n^t \tag{3.29}$$

$$\phi(\mathbf{X}) \leftarrow \mathbf{H}\phi(\mathbf{X})$$

$$\phi(\mathbf{X}_{test}) \leftarrow \phi(\mathbf{X}_{test}) - \frac{1}{n}\mathbf{1}_{n_{test}}\mathbf{1}_n^t\,\phi(\mathbf{X})$$

with $\mathbf{I}$ an n-dimensional identity matrix, and $\mathbf{1}_n$ and $\mathbf{1}_{n_{test}}$ the vectors of ones of dimension n and n_{test}, respectively. Correspondingly, the kernel matrix can be centered in the following way:

$$\mathbf{K} \leftarrow (\mathbf{H}\phi(\mathbf{X}))(\mathbf{H}\phi(\mathbf{X}))^t = \mathbf{HKH} \tag{3.30}$$

$$\mathbf{K}_{test} \leftarrow (\mathbf{H}\varphi(\mathbf{X}))(\varphi(\mathbf{X}_{test}) - \frac{1}{n}\mathbf{1}_{n_{test}}\mathbf{1}_n^t\varphi(\mathbf{X})) = (\mathbf{K}_{test} - \frac{1}{n}\mathbf{1}_{n_{test}}\mathbf{1}_n^t\mathbf{K})\mathbf{H} \quad (3.31)$$

Thus, Equations (3.30) and (3.31) can be further expressed by:

$$\mathbf{K} \leftarrow \mathbf{K} - \mathbf{I}_n\mathbf{K} - \mathbf{K}\mathbf{I}_n + \mathbf{I}_n\mathbf{K}\mathbf{I}_n \quad (3.32)$$

$$\mathbf{K}_{test} \leftarrow \mathbf{K}_{test} - \mathbf{I}_{test}\mathbf{K} - \mathbf{K}_{test}\mathbf{I}_n + \mathbf{I}_{test}\mathbf{K}\mathbf{I}_n \quad (3.33)$$

where $\mathbf{I}_n = (1/n)_{n \times n}$ and $\mathbf{I}_{test} = (1/n)_{n_{test} \times n}$.

3.9   Internet resources

On the World Wide Web you can find many information sources concerning kernel methods and their applications. This section provides general information about these sources.

A comprehensive introduction to kernel methods and support vector machines can be easily found at the Wikipedia website (http://en.wikipedia.org/wiki/Support_vector_machine#Dual_form). The website also provides a number of links related to kernel methods. You can find more detailed knowledge of kernel methods if necessary. Moreover, a formal official website of kernel machines has been established (http://www.kernel-machines.org/). At this website you can find much of the latest progress related to kernel machines, including publications, books, software, tutorials, annual workshops, and so on. In addition, frequently asked questions about kernel machines can be solved through a forum. A support vector machines application list can be found at http://www.clopinet.com/SVM.applications.html. This list covers a great number of applications of the previously proposed support vector machines in many research fields, such as generalized predictive control, dynamic reconstruction of chaotic systems from interspike intervals, the inverse geosounding problem, geo- and environmental sciences, protein fold and remote homology detection, image retrieval, and texture classification, among others. In addition to those mentioned, there are many famous websites on the Web. Two more detailed introductions to all aspects of support vector machines can also be obtained from the websites, http://www.support-vector-machines.org/ and http://www.isis.ecs.soton.ac.uk/resources/svminfo/. We hope that readers interested in kernel methods can get more useful help and information at these websites if necessary.

References

1. Vapnik, V.N. 2000. *The Nature of Statistical Learning Theory*. New York: Springer.
2. Frank, I.E. 1995. Modern nonlinear regression methods. *Chemometrics Intell. Lab.Syst.*, 27:1–19.
3. Breiman, L., Friedman, J.H., Olsen, R.A., and Stone, C.J. (1984) *Classification and Regression Trees*. Wadsworth International, Belmont: CA.
4. Friedman, J.H. 1991. Multivariate adaptive regression splines. *Ann. Statist.*, 19:1–67.
5. Martin, A. 2009. A comparison of nine PLS1 algorithms. *J. Chemometr.*, 23:518–529.
6. Araby, I.A.-R. and Gino, J.L. 2009. A nonlinear partial least squares algorithm using quadratic fuzzy inference system. *J. Chemometr.*, 23:530–537.
7. Walczak, B. and Massart, D.L. 1996. The radial basis functions—Partial least squares approach as a flexible non-linear regression technique. *Anal. Chim. Acta*, 331:177–185.
8. Walczak, B. and Massart, D.L. 1996. Application of radial basis functions—Partial least squares to non-linear pattern recognition problems: Diagnosis of process faults. *Anal. Chim. Acta*, 331:187–193.
9. Wold, S. 1992. Nonlinear partial least squares modelling II. Spline inner relation. *Chemometr. Intell. Lab. Syst.*, 14:71–84.
10. Lombardo, R., Durand, J.F., and Veaux, R.D. 2009. Model building in multivariate additive partial least squares splines via the GCV criterion. *J. Chemometr.*, 23:605–617.
11. Friedman, J.H., Hastie, T., and Tibshirani, R. 2008. New York: Springer.
12. Wythoff, B.J. 1993. Backpropagation neural networks: A tutorial. *Chemometr. Intell. Lab. Syst.*, 18:115–155.
13. Kateman, G. 1993. Neural networks in analytical chemistry? *Chemometr. Intell. Lab. Syst.*, 19:135–142.
14. Svozil, D., Kvasnicka, V., and Pospichal, J. 1997. Introduction to multi-layer feed-forward neural networks. *Chemometr. Intell. Lab. Syst.*, 39:43–62.
15. Mller, K., Mika, S., Rtsch, G., Tsuda, K., and Schlkopf, B. 2001. An introduction to kernel-based learning algorithms. *IEEE Trans. Neural Netw.*, 12:181–202.
16. Shawe-Taylor, J. and Cristianini, N. 2004. Kernel methods for pattern analysis.
17. (1999) *Advances in Kernel Methods: Support Vector Learning*. Cambridge, MA: MIT Press.
18. Smola, A.J. and Schölkopf, B. 2004. A tutorial on support vector regression. *Statist. Comput.*, 14:199–222.
19. Burges, C.J.C. 1998. A tutorial on support vector machines for pattern recognition. *Data Mining Knowl. Discov.*, 2:121–167.
20. Mayoraz, E. and Alpaydin, E. 1999. *Engineering Applications of Bio-Inspired Artificial Neural Networks*, pp. 833–842.
21. Evgeniou, T., Pontil, M., and Poggio, T. 2000. Regularization networks and support vector machines. *Adv. Comput. Math.*, 13:1–50.
22. Bernhard, S., Alexander, S., Klaus-Robert, M., ller. 1998. Nonlinear component analysis as a kernel eigenvalue problem. *Neural Comput.*, 10:1299–1319 DOI http://dx.doi.org/10.1162/089976698300017467.

23. Wold, S. 1987. Principal component analysis. *Chemometr. Intell. Lab. Syst.* 2:37–52.
24. Jolliffe, I.T. 1986. *Principal Component Analysis.* Berlin: Springer.
25. Bernhard, S., lkopf, A.J.S., Klaus-Robert, M., ller (1999) *Advances in Kernel Methods: Support Vector Learning.* Cambridge, MA: MIT Press, pp. 327–352.
26. Wu, W., Massart, D.L., and de Jong, S. 1997. The kernel PCA algorithms for wide data. Part I: Theory and algorithms. *Chemometr. Intell. Lab. Syst.,* 36:165–172.
27. Wu, W., Massart, D.L., and de Jong, S. 1997. Kernel-PCA algorithms for wide data Part II: Fast cross-validation and application in classification of NIR data. *Chemometr. Intell. Lab. Syst.,* 37:271–280.
28. Sebastian, M., Bernhard, S., lkopf, A.S., Klaus-Robert, M., ller, M.S., and Gunnar, R. 1999. tsch *Proceedings of the 1998 Conference on Advances in Neural Information Processing Systems II.* Cambridge, MA: MIT Press.
29. Roman, R. and Leonard, J.T. 2002. Kernel partial least squares regression in reproducing kernel Hilbert space. *J. Mach. Learn. Res.,* 2:97–123.
30. de Jong, S. 1993. SIMPLS: An alternative approach to partial least squares regression. *Chemometr. Intell. Lab. Syst.,* 18:251–263.
31. Rosipal, R. and Krämer, N. 2006. *Subspace, Latent Structure and Feature Selection,* pp. 34–51.
32. Lewi, P.J. 1995. Pattern recognition, reflections from a chemometric point of view. *Chemometr. Intell. Lab. Syst.,* 28:23–33.
33. Keinosuke, F. 1993. *Handbook of Pattern Recognition & Computer Vision.* San Francisco: World Scientific, pp. 33–60.
34. Wu, W., Mallet, Y., Walczak, B., Penninckx, W., Massart, D.L., Heuerding, S., and Erni, F. 1996. Comparison of regularized discriminant analysis linear discriminant analysis and quadratic discriminant analysis applied to NIR data. *Anal. Chim. Acta,* 329:257–265.
35. Hastie, T., Buja, A., and Tibshirani, R. 1995. Penalized discriminant analysis. *Ann. Statist.* 23:73–102.
36. Mika, S., Ratsch, G., Weston, J., Scholkopf, B., and Mullers, K.R. 1999. Neural networks for signal processing IX, 1999. *Proceedings of the 1999 IEEE Signal Processing Society Workshop,* pp. 41–48.
37. Yang, J., Jin, Z., Yang, J.-Y., Zhang, D., and Frangi, A.F. 2004. Essence of kernel Fisher discriminant: KPCA plus LDA. *Pattern Recogn.,* 37:2097–2100.
38. Jian, Y., Frangi, A.F., Jing-Yu, Y., David, Z., and Zhong, J. 2005. KPCA plus LDA: A complete kernel Fisher discriminant framework for feature extraction and recognition. *IEEE Trans. Pattern Anal. Mach. Intell.,* 27:230–244.
39. Baudat, G. and Anouar, F. 2000. Generalized discriminant analysis using a kernel approach. *Neural Comput.,* 12:2385–2404 DOI http://dx.doi.org/10.11 62/089976600300014980.

chapter four

Ensemble learning of support vector machines

Contents

4.1 Introduction

The support vector classifier (SVC) is a linear maximum margin classifier. It can be extended to nonlinear cases by exploiting the idea of a kernel, and transforming the data in the input space to the feature space without explicitly specifying the transformation [1]. Therefore it eventually constructs a linear maximum margin classifier in the feature space. By far, support vector machines, as state-of-the-art methods, perform very well in many classification applications. However, they might still suffer from some practical problems in the training samples [2,3]. For example, when the training set is unbalanced, support vector machines cannot give a good generalization performance. It is worse especially when the given training set has some implicit geometric structure or is not linearly separable in the feature space defined by the chosen kernel.

There are very few theories in the literature to guide us on how to choose kernel functions; the selection of the kernel is usually done in a trial-and-error manner. Thus better results are often difficult and time-consuming to obtain. Furthermore, variance problems may arise when the training set is characterized by very high-dimensional variables and small-sized samples or a large degree of biological variability [3–5]. However, such problems could be addressed through ensemble learning based on support vector machines. The appropriate ensemble learning of support vector machines may be expected to establish a high-performance model with good robustness and reliability. In this chapter, we give a detailed introduction to the ensemble learning of support vector machines in an accessible way. The organization of this chapter is as follows: we start from the idea of ensemble learning, which provides chemists with new important aspects of how to think about chemical and biological problems and build corresponding models. Second, the central concepts about diversity are described in detail, forming the cornerstone of ensemble learning. Finally, two commonly used SVM ensemble methods, boosting in particular, are discussed in detail together with a simple example.

4.2   Ensemble learning

4.2.1   Idea of ensemble learning

As a matter of fact, the general idea of ensemble learning has an intrinsic connection to our daily life experience. We use it all the time! Generally speaking, we hope to improve our confidence in making correct decisions in our daily lives. In fact, making a highly correct decision is certainly a difficult task. How can we address this question in the face of complex problems? We often try to find additional opinions before making a final decision, subsequently weigh various opinions, and combine these opinions through some thought process to reach a final decision [6]. With all opinions, some may gain priority because they come from some experts' decisions. We often place more trust in the decisions of experts because they can provide us with more informative opinions to help us to make correct decisions. However, not all experts can give correct decisions. Once some expert gives a somewhat wrong or imprecise decision, this may bring an even more tremendous risk.

How can we reduce the risk of wrong decisions by some expert? In most cases, combining various opinions by some certain thought process may reduce the risk of an individual opinion that may bring an unfortunate selection. The averaging or weighing combination may or may not exceed the performance of the best opinion in the set of opinions, but it can undoubtedly reduce the overall risk of making a final decision. Ensemble learning just makes full use of such a strategy to improve the individual classifier's performance to a considerably larger extent. Ensemble learning [7–11] aims to build a prediction model by combining the strengths of a collection of simpler basic models. It is made up of two tasks: constructing a population of basic learners from the training data, and then combining them to form the composite predictor. Theoretically speaking, combining several models can effectively reduce the variance of individual models.

Many authors have demonstrated significant performance improvements through ensemble methods. There has been much more literature often referred to as "combining learners" abounding in ad hoc schemes for mixing methods of different types to achieve better performance in chemical fields [12–15]. For example, Xu et al. [16] used partial least squares regression with curds and whey to yield calibration models of high prediction ability. Jiang et al. [17] used a multiple PLS model combination approach to exploit the information comprised in individual models for mid-infrared and near-infrared spectroscopic data. It was disclosed that this ensemble method can avoid possible accumulation of errors in multiple spectral intervals. In fact, many current methods such as bagging [18], boosting [19–25], stacking, consensus modeling method [26], random forest [15, 27–31], artificial neural networks [32], and support vector machines [33] among others are regarded as ensemble learning methods. These methods adopt different strategies to achieve final decision-making with good performance. In the following sections, we mainly focus on the introduction of two representative methods such as bagging and boosting based on support vector machines. However, it is very necessary to dwell on a central concept, that is, diversity of ensemble learning, before we discuss these two representative methods.

4.2.2   Diversity of ensemble learning

Empirically, ensemble methods tend to yield better results when there is significant diversity among the models. Many ensemble methods, therefore, seek to promote diversity among the models they combine. Model diversity formed the cornerstone of ensemble learning [34]. Classifier diversity can be achieved in several ways. The popular method is to use different training datasets to train individual classifiers. Such datasets are often obtained through resampling techniques with or without replacement. Examples of this approach include bagging, boosting, and their many variants [35–38]. For example, the model diversity of bagging may be obtained by using different training data to independently train individual learners.

Another approach to achieve diversity is to use different training parameters for different classifiers. For example, a series of multilayer perceptron neural networks can be trained using different weight initializations, the number of layers/nodes, error goals, and so on; adjusting these parameters allows one to control the instability of the individual classifiers, and hence contribute to their diversity. The ability to control the instability of neural networks makes the neural network a suitable candidate to be used in ensemble learning methods. Alternatively, an entirely different type of classifier such as the k-nearest neighbor method, support vector machines, and decision trees can also be combined for added diversity. However, combining different models, even different architectures of the same model, is only used for the specific application that warrants it. Finally, diversity can also be achieved by using different feature subsets. They can be randomly drawn from the entire feature pool, and hence construct different classifiers. Such a method is usually called the random subspace method [39–41]. The most notable example of this approach is the random forest method, which uses decision trees as basic models.

4.3   Bagging support vector machines

Bagging [18], short for bootstrap aggregating, is a bootstrap ensemble method that creates individuals for its ensemble by training each classifier on a random redistribution of the training set, thus incorporating the benefits of both the bootstrap and aggregating approaches. Figure 4.1 shows the bagging procedure scheme. It can clearly be seen from Figure 4.1 that many bootstrap samples are first generated from the original training set. Bootstrap [42] is based on random sampling with replacement. Consider the regression problem: suppose we want to fit a SVM model to our training data $\mathbf{Z}$, obtaining the prediction $f(\mathbf{x})$ at input $\mathbf{x}$. First, N samples are generated by randomly drawing with replacement, where $\mathbf{Z}^* = \{(\mathbf{x}^*_1, y^*_1), (\mathbf{x}^*_2, y^*_2), ..., (\mathbf{x}^*_N, y^*_N)\}$. It is worth noting that

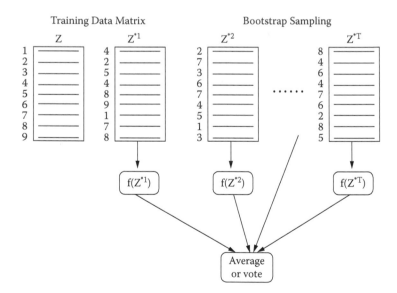

Figure 4.1 The schematic illustration of bagging procedure. *m*: number of objects. *T*: number of iterations. Average *T* predicted values for regression; cast a vote of *T* predicted values for classification.

many of the original samples may be repeated in the resulting training set whereas others may be left out (approximately $1/e \approx 37\%$ of all the samples are not presented in the bootstrap sample). The main idea of aggregating actually means combining multiple models. For each bootstrap set $\mathbf{Z}^{*t}$, $t = 1, 2, \dots, T$, one model is fitted, giving prediction $f(\mathbf{x})$. The estimation of bagging SVM, averaging the predictions over a collection of bootstrap observations, is given by:

$$f_{bag}(x) = \frac{1}{T} \sum_{t=1}^{T} f^{*t}(x) \tag{4.1}$$

It is noticeable that combining models never makes the root mean square of residuals worse at the population level, because it can dramatically reduce the variance of individual algorithms with the help of averaging.

4.4   Boosting support vector machines

4.4.1   Boosting: A simple example

Boosting is a general and effective method for producing an accurate prediction rule by combining rough and moderately rough rules of thumb. Given a classification problem, the goal, of course, is to generate a rule that makes the most accurate predictions possible on new test samples. To apply boosting, there are two fundamental questions that must be solved: how should each weak classifier be chosen, and once we have collected many weak classifiers, how should the weak classifiers be combined into a single one? Subsequently, we show how boosting solves the above questions mathematically.

In order briefly to elucidate how the boosting technique works, let's consider a binary classification problem in 2-D space. Let the ith sample be denoted by $\mathbf{x}_i = [x_{i1}, x_{i2}]$. The data were labeled by two types of marks that differentiate two classes. Each sample belongs to either +1 (circle) or −1 (triangle) and cannot be separated linearly (see Figure 4.2). We choose the split along either x_1 or x_2 that produces the largest decrease in training misclassification error as our classifier stump. Figure 4.3 shows the overall process of the AdaBoost algorithm. Initially all of the weights were set to $\omega_{1i} = 1/10$ ($i = 1, 2, \dots, 10$), so that the first step simply trains the classifier in the usual way. The line (h_1) along the x_2-direction splits the samples into two parts, and two samples (solid black triangle) are misclassified. After the first iteration, the misclassified samples are given a high weight ($\omega_{2i} = 0.25$ for misclassified samples, $\omega_{2i} = 0.0625$ for correctly classified samples) so that the next classifier could focus on these misclassified

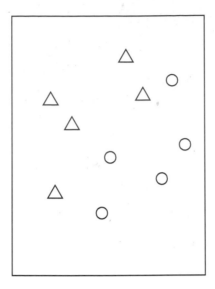

Figure 4.2 The simulated data for two class problems marked by +1 (circle) and −1 (triangle).

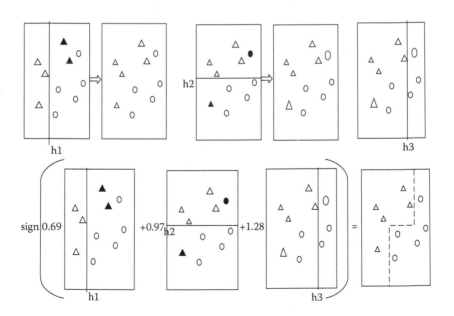

Figure 4.3 The overall iteration process for AdaBoost.

samples. In order to emphasize this, the samples with larger signs indicate their misclassification with larger weights in the plot.

To achieve the minimal weighted misclassification error rate, the adjustment of weight distribution forces the next classifier to correctly classify the samples with larger weights. Thus, the line (h_2) along the x_1-direction splits the weighted samples into two parts, in which there are still two samples (solid black circle) being misclassified with higher weights (ω_{3i} = 0.25 for samples misclassified in this step; ω_{3i} = 0.143 for samples misclassified in the first step; ω_{3i} = 0.0375 for samples all correctly classified in the above two steps). With the help of these weights, the third split (h_3) can occur along the x_2-direction on the far-right side (see Figure 4.3). It is worthwhile noting that as iterations proceed, those samples misclassified by the classifier induced at the previous step have their weights increased, whereas the weights are decreased for those classified correctly. Finally, the predictions from all of them are then combined through a weighted majority vote to produce the final prediction:

$$G(\mathbf{x}) = sign(\sum_{t=1}^{T} c_t G_t(\mathbf{x})) \qquad (4.2)$$

Here c_1, c_2, ..., c_t are computed by the boosting algorithm and weigh the contribution of each respective classifier $G(\mathbf{x})$. Their effect is to give more influence to the more accurate classifiers in the sequence. In the example, the weights of the three classifiers are 0.69, 0.97, and 1.28, respectively. Figure 4.4 shows the boosting procedure for classification. In the plot, it can clearly be seen that the AdaBoost algorithm calls this weak classifier repeatedly, each time providing it with training examples with different weight distribution. After many rounds, the boosting algorithm must combine these weak rules into a single prediction rule that, it is hoped, will be much more accurate than any one of the weak rules.

4.4.2 Boosting SVMs for classification

The commonly used boosting method for classification problems in the chemistry and biotechnology fields is Freund's AdaBoost. There are three types of AdaBoost algorithms: discrete AdaBoost, real AdaBoost, and gentle AdaBoost. In addition, the AdaBoost.MH algorithm can be applied for multiclass classification. The algorithms are described in detail in References [13,21]. The discrete AdaBoost algorithm as a representative example is discussed in this section.

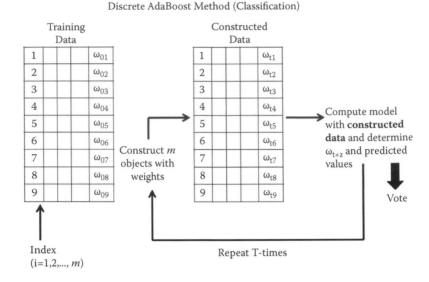

Figure 4.4 The schematic illustration of boosting procedure for classification. *m*: number of objects. *T*: number of iterations. ω_{ti}: the weight of the *i*th object at the *t*th iteration. Vote: cast a vote of *T* predicted values for classification.

Consider a training dataset with *N* samples belonging to two classes. The two classes are labeled $y \in \{-1, 1\}$. The discrete AdaBoost algorithm based on SVM consists of the following steps:

1. Assign initial equal weights to each sample in the original training set:

$$w_i^1 = 1/N, i = 1, 2, ..., N$$

2. For iterations $t = 1, 2, ..., T$:
   (a) Select a dataset with N samples from the original training set using weighted resampling. The chance for a sample to be selected is related to its weight. A sample with a higher weight has a higher probability to be selected.
   (b) Train a SVC $f(\mathbf{x})$ from the resampled dataset.
   (c) Apply the SVC $f(\mathbf{x})$ to the original training dataset. If a sample is misclassified, its error $err = 1$, otherwise its error is 0.
   (d) Compute the sum of the weighted errors of all training samples.

$$err^t = \sum_{i=1}^{N} (w_i^t \times err_i^t) \qquad (4.3)$$

3. The confidence index of the SVC $f(\mathbf{x})$ is calculated as:

$$c^t = \log((1 - err^t) / err^t) \qquad (4.4)$$

The lower the weighted error made by the SVC $f(\mathbf{x})$ on the training samples is, the higher the confidence index of the SVC $f(\mathbf{x})$ is.

(e) Update the weight of all original training samples.

$$w_i^{t+1} = w_i^t \exp(c^t err_i^t) \qquad (4.5)$$

The weights of the samples that are correctly classified are unchanged and the weights of the misclassified samples are increased.

(f) Renormalize w so that $\displaystyle\sum_{I=1}^{N} W_I^{t+1} = 1$

(g) $T = t + 1$, if err <0.5 and $t < T$, repeat steps (a)–(f); otherwise, stop and $T = t - 1$. After T iterations in Step 2, there are T SVC $f^t(\mathbf{x})$, $t = 1, 2, \ldots, T$.

The performance of discrete AdaBoost is evaluated by a test set. For a sample j of the test set, the final prediction is the combined prediction obtained from the T learners. Each prediction is multiplied by the confidence index of the corresponding SVM classifier $f(\mathbf{x})$. The higher the confidence index of a SVC $f(\mathbf{x})$ is, the higher is its role in the final decision.

$$y_j = sign(\sum_{t=1}^{T} c^t f^t(\mathbf{x}_j)) \qquad (4.6)$$

In the algorithm, several interesting features are especially worth noting. The AdaBoost algorithm generates a set of SVCs, and combines them through weighted majority voting of the classes predicted by the individual SVC. Each SVC is obtained using samples drawn from an iteratively updated distribution of all the training samples. Thus, AdaBoost maintains a distribution or set of weights over the original training set $\mathbf{Z}$ and adjusts these weights after each SVM classifier is learned. The adjustments increase the weight of samples that are misclassified by SVM classifiers and decrease the weight of samples that are correctly classified. Hence, consecutive classifiers gradually focus on those increasingly hard to classify samples.

Equation (4.4) describes the confidence index of the SVM classifier. The higher the confidence index is, the more accurate its corresponding SVM classifier is. A large confidence index indicates that the corresponding SVM classifier plays a more important role in the final decision. AdaBoost adopts such a strategy called weighted majority voting to weigh a set of classifiers. The idea behind the strategy is very intuitive and effective. Those classifiers that have shown high performance during training are rewarded with higher voting weights than others. Note that the confidence index is larger than 0. This guarantees that the SVC is better than random guessing. Equation (4.5) describes the training set distribution update rule. Each distribution update of the training set is controlled by the weights of the training samples. The distribution weights of the samples that are correctly classified are unchanged whereas the distribution weights of the misclassified samples are increased. Thus, iteration by iteration, AdaBoost updates the distribution of the training samples in order to give more attention toward those increasingly difficult samples. Once the preset number of iterations is reached, AdaBoost is ready to classify the unlabeled test samples.

Practically, AdaBoost has many advantages. It is fast, simple, and easy to program. It has no parameters to tune (except for the number of iterations T). Moreover, it also requires no prior knowledge about the classifier and can flexibly be implemented with any classification method. A nice property of AdaBoost is its ability to identify outliers [43], that is, samples that are either mislabeled in the training data, or are inherently ambiguous and hard to classify. Because AdaBoost focuses on the samples that are more difficult to classify, the samples with the highest weights often turn out to be outliers.

4.4.3   Boosting SVMs for regression

There are two methods used to establish the boosting SVR model. The first one is by forward stagewise additive modeling, which modifies the target values to effectively fit residuals. The second one is by imitating classification and changing sample weights to emphasize those that were poorly regressed in previous stages of the fitting process [44,45]. Inasmuch as the latter is similar to boosting for classification, we mainly focus on the forward stagewise additive model in this section.

Generally speaking, an additive model can be expressed as

$$F(\mathbf{x}) = \beta_1 f_1(\mathbf{x}) + \beta_2 f_2(\mathbf{x}) + \ldots + \beta_T f_T(\mathbf{x}) = \sum_{t=1}^{T} \beta_t f_t(\mathbf{x}) \qquad (4.7)$$

where β_t, $t = 1, 2, \ldots, T$ are the expansion coefficients, typically $\beta_t = 1$, and the functions $f_t(x)$ are usually chosen as simple functions with input x. The boosting method aims at finding a latent function $F(x)$ that minimizes a loss function such as squared error loss, and the like between response y and estimated values $F(x)$ (in most cases, the latent function is not known to us). So, it attempts to seek a simple function $f(x)$ at each iteration.

$$\min_{f(x)} \sum_{i=1}^{N} \| y_i - F_{t-1}(x_i) - f_t(x_i) \|^2 \tag{4.8}$$

where t is the number of the $f(x)$ added. Thus, each simple function can be generated by fitting previous residuals with the original training set X.

The algorithm of boosting SVR based on the forward stagewise additive strategy is the following:

1. Initially fit the training data using a SVR model $f(X)$, expressed as $\hat{y}_1 = f_1(X)$.

   Calculate the residual:

   $$\mathbf{y}_{res} = \mathbf{y} - v_1 \hat{\mathbf{y}}_1 \tag{4.9}$$

   Here $0 < v < 1$; v is a shrinkage parameter that may or may not be a constant. This shrinkage parameter can effectively prevent overfitting problems. Only the v part of the fitted values is extracted at each step.
2. For $t = 2, \ldots, T$, repeat the following steps:
   (a) Fit the current residual $\mathbf{y}_{res/t-1}$ using SVR model: $\hat{y}_t = f_t(X)$.
   (b) Update the residual,

   $$\mathbf{y}_{res/t} = \mathbf{y}_{res/t-1} - v_t \hat{\mathbf{y}}_t \tag{4.10}$$

   In this step, only the v part of the fitted $\hat{y}_t$ is used as useful regression information.
3. The final prediction is

$$\mathbf{y}_{pre} = v_1 \hat{\mathbf{y}}_1 + v_2 \hat{\mathbf{y}}_2 + \cdots + v_{T-1} \hat{\mathbf{y}}_{T-1} + v_T \hat{\mathbf{y}}_T = \sum_{t=1}^{T} v_t f_t(X) \tag{4.11}$$

The basic idea of boosting SVR as an additive model is to sequentially construct additive regression models by fitting a basic SVR model to the current residuals that are not fitted by previous models. The current residuals are the gradient of the loss function (e.g., a squared error function) being minimized at the current step. Initially, a basic SVR model is generated by fitting the response y with the predictor X, and the initial residual of **y** is calculated. Hence, iteration by iteration, boosting only considers the current residual of **y** and always fits the current residual of **y** with the original **X**. Thus, boosting sequentially adds a series of basic SVR models that are simple to implement. Each constructed model is shrunk by a shrinkage parameter v, which is a positive value less than 1. The v acts as a weight for the current model that prevents possible overfitting by constraining the fitting process [46]. The whole **X** is used each time to extract information correlated to the residuals of **y**, but the information extracted is not completely used. Only the v part is used, and the rest, the $1 - v$ part, is put back to avoid overfitting. The sequential adding is repeated, resulting in many weighted basic SVR models. Once a preset number T of iterations is reached, a single test set will be used for prediction. The final prediction is the combination of these weighted basic regression models.

Instead of the above squared error loss function, other robust loss functions with accompanying gradient functions can also be implemented. Table 4.1 summarizes the gradients of commonly used loss functions. For squared error loss, the negative gradient is just the ordinary residual, as pointed out above. For absolute error loss, the negative gradient is the sign of the residual. Thus, the signs of the current residuals are fitted with the original predictor **X** at the current step. For Huber M-regression, the negative gradient is the combination between the above two (see Table 4.1).

4.4.4   Further consideration

Some key problems related to boosting should be worth noting, which may be very helpful to our in-depth understanding of the boosting technique. (1) In general, good performance of ensemble learning can be expected when the individual classifiers are diverse. For boosting, model diversity may be obtained by using different training data to sequentially train individual learners. Such datasets are often generated through weighted resampling techniques where training data subsets are drawn randomly with probability proportional to their weights. There are two ways that AdaBoost can use these weights to construct a new training set **Z*** to enhance a weak learner. (a) Boosting by sampling: samples are drawn with replacement from **Z** with probability proportional to their weights. Samples with high

Table 4.1 Gradients for Commonly Used Loss Functions

	Loss Function	Gradient
Squared error loss	$1/2\,[y_i - F(x_i)]^2$	$y_i - F(x_i)$
Absolute error loss	$y_i - F(x_i)$	$\text{sign}\,[y_i - F(x_i)]$
Huber	$1/2\,[y_i - F(x_i)]^2$ for $y_i - F(x_i) \leq \delta$	$y_i - F(x_i)$ for $y_i - F(x_i) \leq \delta$
	$(y_i - F(x_i)$ for $y_i - F(x_i) > \delta$	$\delta(y_i - F(x_i))$ for $y_i - F(x_i) > \delta$

Note: δ is the αth quantile $\{y_i - F(x_i)\}$.

probability will occur more often than those with low probability, and some samples may not occur in the constructed training samples at all although their probability is not zero. (b) Boosting by weighting: weak learners can accept a weighted training set directly. With such an algorithm, the entire training set **Z** with associated weights is given to the weak learner. Both methods have been shown to be very effective.

(2) A crucial property of boosting is that it appears to be more resistant to overfitting. Recently, Massart used a boosting algorithm combined with PLS (BPLS) to predict seven chemical datasets and showed that BPLS is relatively resistant to overfitting [47]. Overfitting is a commonly observed situation where the learner performs well on training data, but has a large error on the test data [48]. Overfitting is usually attributed to memorizing the data, or learning the noise in the data. Much to our surprise, boosting can continuously decrease the generalization error even after training error reaches zero. Schapire et al. later provided an explanation to this phenomenon based on the so-called margin theory. The details of margin theory can be found in Reference [49]. Generally speaking, the margin of a sample **x** can be considered as its distance from the decision boundary. The farther away a sample is from the boundary, the larger its margin, and hence the higher the confidence of the classifier in correctly classifying this sample. Similar to the margin of SVMs, Schapire et al. used a slightly different definition of a margin for AdaBoost and showed that AdaBoost also boosts the margins. That is, it can find the decision boundary having maximum distances to all samples. In the context of AdaBoost, the margin of a sample is simply the difference between the total votes it receives from correctly identifying classifiers and the maximum vote received by any incorrect class. They show that the ensemble error is bounded with respect to the margin, but is independent of the number of classifiers. The concept of a margin for AdaBoost and SVMs is conceptually illustrated in Figure 4.5.

(3) Bagging and boosting being non-Bayesian methods have, to a certain extent, some similarity with the Bayesian model [50]. A general form for bagging, boosting, and Bayesian models can be expressed as:

$$\hat{F}(\mathbf{x}) = \sum_{t=1}^{T} w_t E(\mathbf{y} \mid \mathbf{x}, \hat{\theta}_t) \tag{4.12}$$

where $\hat{\theta}_t$ is a set of submodel parameters and w_t the weight of the tth model. $E(*)$ is the expectation value at some point. For Bayesian models the $w_t = 1/T$, and the average estimates the posterior mean by sampling $\hat{\theta}_t$ from the posterior distribution. That is, the Bayesian approach fixes the data and perturbs the parameters, according to the current estimate of the posterior distribution. For bagging, $w_t = 1/T$ and $\hat{\theta}_t$ is the parameter vector refit to bootstrap samples of the training data. That is, bagging perturbs the data

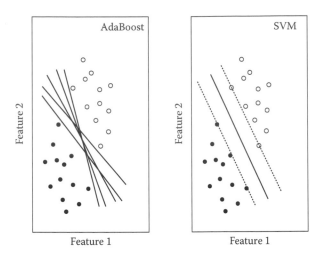

Figure 4.5 Both SVMs and AdaBoost maximize the margin between samples of different classes.

in an i.i.d. (independent and identically distributed) fashion and then re-estimates the model to give a new set of model parameters. At the end, a simple average of the model predictions from different bagged samples is computed. For boosting, all weights are equal to 1, but θ_t is typically selected in a nonrandom sequential fashion to constantly improve the fit. That is, boosting fits a model that is additive in the models of each individual basic learner, which are learned using non-i.i.d. samples. The comparison of the three methods can help us gain a deeper understanding of the boosting concept.

(4) Compared with bagging, boosting has a significant advantage. Boosting can not only reduce the variance of the weak learner by averaging several weak learners obtained from different subsamples of the training samples, but also reduce the bias of the weak learner by forcing the weak learner to concentrate on hard-to-classify samples [51,52].

(5) The underlying idea of booting is deeply rooted in our personal daily decision making. It provides chemists with new important aspects of how to think about chemical and biological problems and build corresponding models. In practice, certain chemical or biological problems are just too difficult for a given individual learner to solve. More specifically, the form of the target (modeled) function may be too complex to model with a single model, or the target function may lie outside the selected model function space (wrong model). Typical examples in chemistry are QSAR/QSPR studies [53]. Because of the diversity of chemicals, the truth may be a very complex function form (nonlinear) and it seems therefore hard to construct an accurate integral QSAR/QSPR model generalizing the

diverse datasets. However, finding many relatively linear functions can be a lot easier than finding a single, highly accurate nonlinear function. Thus, such questions might resort to sequentially combining appropriate simple linear functions to approximate the latent nonlinear function.

References

1. Vapnik, V. N., *The nature of statistical learning theory.* New York: Springer-Verlag 2000.
2. Li, G.-Z.; Meng, H.-H.; Lu, W.-C.; Yang, J.; Yang, M., Asymmetric bagging and feature selection for activities prediction of drug molecules. *BMC Bioinformatics* **2008**, 9, S7.
3. Valentini, G.; Muselli, M.; Ruffino, F. In *Bagged ensembles of Support Vector Machines for gene expression data analysis*, Neural Networks, 2003. Proceedings of the International Joint Conference on, 2003; 2003; pp 1844–1849 vol.3.
4. Zhou, Y.-P.; Jiang, J.-H., Boosting support vector regression in QSAR studies of bioactivities of chemical compounds. *European Journal of Pharmaceutical Sciences* **2006**, 28, 344–353.
5. Li, Y.; Shao, X.; Cai, W., A consensus least squares support vector regression (LS-SVR) for analysis of near-infrared spectra of plant samples. *Talanta* **2007**, 72, 217–222.
6. Polikar, R., Ensemble Based Systems in Decision Making *IEEE Circuits and Systems Magazine* **2006**, 6, 21–45.
7. Breiman, L., Arcing classifiers. *Annals of Statistics* **1998**, 26, 801–824.
8. Breiman, L., Prediction games and arcing algorithms. *Neural Computation* **1999**, 11, 1493–1517.
9. Breiman, L., Randomizing outputs to increase prediction accuracy. *Machine Learning* **2000**, 40, 229–242.
10. Schapire, R. E. In *Theoretical views of boosting and applications*, 10th International Conference on Algorithmic Learning Theory (ALT 99), Tokyo, Japan, Dec 06–08, 1999; Watanabe, O. Y. T., Ed. Tokyo, Japan, 1999; pp 13–25.
11. Arodz, T.; Yuen, D. A.; Dudek, A. Z., Ensemble of Linear Models for Predicting Drug Properties. *Journal of Chemical Information and Modeling* **2006**, 46, 416–423.
12. Xu, Q. S.; Daszykowski, M.; Walczak, B.; Daeyaert, F.; de Jonge, M. R.; Heeres, J.; Koymans, L. M. H.; Lewi, P. J.; Vinkers, H. M.; Janssen, P. A.; Massart, D. L., Multivariate adaptive regression splines–studies of HIV reverse transcriptase inhibitors. *Chemometrics and Intelligent Laboratory Systems* **2004**, 72, 27–34.
13. Zhang, M. H.; Xu, Q. S.; Daeyaert, F.; Lewi, P. J.; Massart, D. L., Application of boosting to classification problems in chemometrics. *Analytica Chimica Acta* **2005**, 544, 167–176.
14. Svetnik, V.; Wang, T.; Tong, C.; Liaw, A.; Sheridan, R. P.; Song, Q., Boosting: An Ensemble Learning Tool for Compound Classification and QSAR Modeling. *Journal of Chemical Information and Modeling* **2005**, 45, 786–799.
15. Svetnik, V.; Liaw, A.; Tong, C.; Culberson, J. C.; Sheridan, R. P.; Feuston, B. P., Random Forest: A Classification and Regression Tool for Compound Classification and QSAR Modeling. *Journal of Chemical Information and Computer Sciences* **2003**, 43, 1947–1958.

16. Xu, Q. S.; de Jong, S.; Lewi, P.; Massart, D. L., Partial least squares regression with Curds and Whey. *Chemometrics and Intelligent Laboratory Systems* **2004,** 71, 21–31.

17. Jiang, J. H.; Berry, R. J.; Siesler, H. W.; Ozaki, Y., Wavelength interval selection in multicomponent spectral analysis by moving window partial least-squares regression with applications to mid-infrared and hear-infrared spectroscopic data. *Analytical Chemistry* **2002,** 74, 3555–3565.

18. Breiman, L., Bagging predictors. *Machine Learning* **1996,** 24, 123–140.

19. Freund, Y., Boosting a weak learning algorithm by majority. *Information and Computation* **1995,** 12, 252–285.

20. Freund, Y. In *An adaptive version of the boost by majority algorithm*, 12th Annual Conference on Computational Learning Theory, Santa Cruz, California, Jul 07–09, 1999; Santa Cruz, California, 1999; pp 293–318.

21. Friedman, J. H.; Hastie, T.; Tibshirani, R., Additive logistic regression: A statistical view of boosting. *Annals of Statistics* **2000,** 28, 337–374.

22. Friedman, J. H., Greedy function approximation: A gradient boosting machine. *Annals of Statistics* **2001,** 29, 1189–1232.

23. Friedman, J. H., Stochastic gradient boosting. *Computational Statistics & Data Analysis* **2002,** 38, 367–378.

24. Freund, Y.; Iyer, R.; Schapire, R. E.; Singer, Y., An efficient boosting algorithm for combining preferences. *Journal of Machine Learning Research* **2004,** 4, 933–969.

25. Gey, S.; Poggi, J.-M., Boosting and instability for regression trees. *Computational Statistics & Data Analysis* **2006,** 50, 533–550.

26. Li, J. Z.; Lei, R. L.; Liu, H. X.; Li, S. Y.; Yao, X. J.; Liu, M. C.; Gramatica, P., QSAR Study of Malonyl-CoA Decarboxylase Inhibitors Using GA-MLR and a New Strategy of Consensus Modeling. *Journal of Computational Chemistry* **2008,** 29, 2636–2647.

27. Breiman, L., Random forests. *Machine Learning* **2001,** 45, 5–32.

28. Palmer, D. S.; O'Boyle, N. M.; Glen, R. C.; Mitchell, J. B. O., Random Forest Models To Predict Aqueous Solubility. *Journal of Chemical Information and Modeling* **2007,** 47, 150–158.

29. Nandigam, R. K.; Evans, D. A.; Erickson, J. A.; Kim, S.; Sutherland, J. J., Predicting the Accuracy of Ligand Overlay Methods with Random Forest Models. *Journal of Chemical Information and Modeling* **2008,** 48, 2386–2394.

30. Li, S.; Fedorowicz, A.; Singh, H.; Soderholm, S. C., Application of the Random Forest Method in Studies of Local Lymph Node Assay Based Skin Sensitization Data. *Journal of Chemical Information and Modeling* **2005,** 45, 952–964.

31. Ehrman, T. M.; Barlow, D. J.; Hylands, P. J., Virtual Screening of Chinese Herbs with Random Forest. *Journal of Chemical Information and Modeling* **2007,** 47, 264–278.

32. Agrafiotis, D. K.; Cedeno, W.; Lobanov, V. S., On the Use of Neural Network Ensembles in QSAR and QSPR. *Journal of Chemical Information and Computer Sciences* **2002,** 42, 903–911.

33. Li, H.; Liang, Y.; Xu, Q., Support vector machines and its applications in chemistry. *Chemometrics and Intelligent Laboratory Systems* **2009,** 95, 188–198.

34. Kuncheva, L. I.; Whitaker, C. J., Measures of Diversity in Classifier Ensembles and Their Relationship with the Ensemble Accuracy. *Machine Learning* **2003,** 51, 181–207.

35. Rudin, C.; Schapire, R. E.; Daubechies, I. In *Boosting based on a smooth margin,* 17th Annual Conference on Learning Theory (COLT 2004), Banff, Canada, Jul 01–04, 2004; Shawe-Taylor, J. S. Y., Ed. Banff, Canada, 2004; pp 502–517.
36. Rudin, C.; Schapire, R. E.; Daubechies, I., Analysis of boosting algorithms using the smooth margin function. *Annals of Statistics* **2007**, 35, 2723–2768.
37. Shafik, N.; Tutz, G., Boosting nonlinear additive autoregressive time series. *Computational Statistics & Data Analysis* **2009**, 53, 2453–2464.
38. Hong, P. Y.; Liu, X. S.; lu, X.; Liu, J. S.; Wong, W. H., A boosting approach for motif modeling using ChIP-chip data. *bioinformatics* **2005**, 21, 2636–2643.
39. Skurichina, M.; Duin, R., Bagging and the Random Subspace Method for Redundant Feature Spaces. In *Multiple Classifier Systems*, 2001; pp 1–10.
40. Tin Kam, H., The Random Subspace Method for Constructing Decision Forests. *IEEE Trans. Pattern Anal. Mach. Intell.* **1998**, 20, 832–844.
41. Skurichina, M.; Duin, R. P. W., Bagging, Boosting and the Random Subspace Method for Linear Classifiers. *Pattern Analysis & Applications* **2002**, 5, 121–135.
42. Efron, B.; Tibshirani, R. J., *An Introduction to the Bootstrap*. Chapman & Hall: 1993.
43. Cheze, N.; Poggi, J. M., Outlier detection by boosting regression trees. *Journal of Statistical Research of Iran* **2006**, 3, 1–21.
44. Freund, Y.; Schapire, R. E., A decision-theoretic generalization of on-line learning and an application to boosting. *Journal of Computer and System Sciences* **1997**, 55, 119–139.
45. Drucker, H., Improving Regressors using Boosting Techniques. In *machine learning-international workshop then conference*, 1997.
46. Copas, J. B., Regression, Prediction and Shrinkage. *Journal of the Royal Statistical Society. Series B (Methodological)* **1983**, 45, 311–354.
47. Zhang, M. H.; Xu, Q. S.; Massart, D. L., Boosting Partial Least Squares. *Analytical Chemistry* **2005**, 77, 1423–1431.
48. Hawkins, D. M., The Problem of Overfitting. *Journal of Chemical Information and Computer Sciences* **2004**, 44, 1–12.
49. Schapire, R. E.; Freund, Y.; Bartlett, P.; Lee, W. S., Boosting the margin: A new explanation for the effectiveness of voting methods. *Annals of Statistics* **1998**, 26, 1651–1686.
50. Friedman, J. H.; Hastie, T.; Tibshirani, R., The Elements of Statistical Learning: Data Mining, Inference and Prediction. In Springer-Verlag: New York, 2008.
51. Freund, Y.; Schapire, R. E. In *Experiments with a New Boosting Algorithm,* Machine Learning: Proceedings of the Thirteenth International Conference, 1996; 1996.
52. Kong, E. B.; Dietterich, T. In *Error-correcting output coding corrects bias and variance* In Proceedings of the Twelfth International Conference on Machinde Learning, 1995; 1995; pp 313–321.
53. Cao, D.-S.; Liang, Y.-Z.; Xu, Q.-S.; Li, H.-D.; Chen, X., A new strategy of outlier detection for QSAR/QSPR. *Journal of Computational Chemistry* **2010**, 31, 592–602.

chapter five

Support vector machines applied to near-infrared spectroscopy

Contents

5.1 Introduction

In Chapters 1 and 2, the basic elements of SVM are described in detail. Also presented there are the results of both classification and regression simulation studies. The kernel function used in SVMs offers an alternative solution by projecting the data into a high-dimensional feature space to increase computational power. The linearly inseparable data in the original low-dimensional space could become linearly separable after being projected into the high-dimensional feature space. Thus, the kernel method is discussed thoroughly in Chapter 3 to exploit the nonlinear ability of SVMs. In recent years, ensemble learning as a new and promising learning strategy, say bagging and boosting, has been gaining increasing attention and application in many fields. In this context, the ensemble of SVMs was therefore introduced in Chapter 4 and its performance tested on some datasets. On the whole, the first four chapters are mainly focused on the theoretical aspects of SVM. Having an understanding of the mechanics of SVM, one would like to investigate the performance of

SVMs when dealing with real data. In this chapter, we concentrate on the applications of SVMs to near-infrared data.

5.2   Near-infrared spectroscopy

In mysterious nature, there are many interesting phenomena related to light, such as the colorful rainbow and the wonderful mirage and so on. Maybe all these could be ascribed to the interaction of light with matter, which has captured people's interest for the last two millennia. As early as 130 A.D., Ptolemaeus tabulated the refraction of light for a range of transparent materials, and in 1305, Von Freiburg simulated the structure of the rainbow by using water-filled glass spheres [1]. With the passing of time, much has been learned regarding light, such as the laws of reflection and refraction, through the work of great scientists such as Snell, Newton, and Priestley. During the nineteenth century, near-infrared (NIR) radiation was discovered by Herschel, who was a successful musician turned astronomer. By the beginning of the twentieth century, the nature of the electromagnetic spectrum was much better understood than before. However, little progress outside the visible part of the spectrum was achieved due to the lack of suitable detection equipment.

The early 1880s saw an important step in understanding the electromagnetic spectrum due to the work by Abney and Festing who recorded the spectra of organic liquids in the range 1 to 1.2 μm in 1881 [1]. This work marked great progress because (1) it represented the first serious NIR measurements and (2) Abney and Festing recognized the importance of the hydrogen bond in the NIR spectrum. Following this work, Coblentz constructed a very susceptible spectrometer by which the spectra of several hundred compounds were measured, although the method was very time-consuming.

A few decades later, in the 1950s, the first industrial application of NIR began. In the first applications, near-infrared spectroscopy (NIRS) was used only as an add-on unit to other optical devices that used other wavelength bands such as ultraviolet (UV), visible (Vis), or mid-infrared (MIR) spectrometers. In the 1980s, a single-unit, stand-alone NIRS system was made available, but the application of NIRS was focused more on chemical analysis. With the introduction of light-fiber optics in the mid-1980s and the monochromator detector developments in early 1990s, NIRS became a more powerful tool for scientific research and practical applications. A schematic of NIR spectroscopy is shown in Figure 5.1 (cited from [2]). Typical applications of NIR spectroscopy include, but are not limited to, pharmaceutical [3–5], medical diagnostics (including blood sugar and oximetry) [6], food chemistry [7–11], and agricultural products [12–15] among others. Specifically, we would like to introduce here a great scientist in NIR analysis, Professor Karl Norris. He was a major force in the development of near-infrared reflection technology for the rapid, invasive,

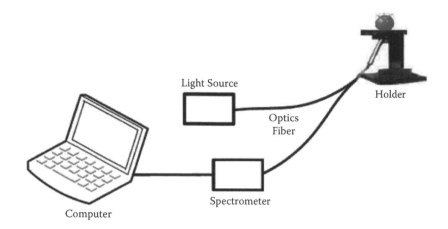

Figure 5.1 A schematic of the spectral measurement setup.

and inexpensive testing of many quality characteristics of agricultural products and food. Due to his prominent contributions, he is regarded as the "father" of modern near-infrared spectroscopic analysis [1].

Generally speaking, near-infrared spectroscopy is a spectroscopic technology that utilizes the near-infrared region of the electromagnetic spectrum (from about 800 nm to 2500 nm). NIR spectroscopy is typically used for quantitative measurement of organic functional groups, especially O–H, N–H, and C–H. Detection limits are typically 0.1% and applications include pharmaceutical, agricultural, polymer, and clinical analysis.

Let's now put NIR spectroscopy in a slightly theoretical context. It is based on molecular overtones and combination vibrations. The molecular overtone and combination bands seen in the near-IR region are typically very broad, leading to a very complex spectrum. For example, any compound containing a C–H, N–H, or O–H bond will have a contribution to the NIR spectrum of a given sample, such as an apple. For this reason, it can be difficult to assign a specific wavelength to specific chemical components. So, near-infrared spectroscopy itself is not a particularly sensitive technique. In other words, the chemical information contained in an NIR spectrum is not easily extracted or understood.

To a large extent, the successful applications of NIR spectroscopy should be ascribed to the multivariate (multiple wavelength) calibration techniques, which can be divided into two categories. The former refers to linear calibration methods, for example, ridge regression (RR) [16], principal components regression (PCR) [17, 18], and partial least squares (PLS) [19–22], and the latter includes artificial neural networks (ANNs) [23–25], support vector machines (SVMs) [26–32], and the like, which are much more flexible in correlating the quality index under consideration, such as moisture content, with the digitalized NIR spectra.

5.3 Support vector machines for classification of near-infrared data

In this section, we begin by analyzing an NIR dataset measured in our laboratory by using support vector classification (SVC) machines. The performance of SVC is compared to that of partial least squares-linear discriminant analysis (PLS-LDA), which is a frequently used classification method in chemometrics. In addition, the performance of SVC and PLS-LDA after variable wavelength selection is also presented. Next, some related work from other groups is presented, aiming at providing more examples for interested readers.

5.3.1 Recognition of blended vinegar based on near-infrared spectroscopy

NIR spectra were collected using an Antaris ‖ FT-NIR spectrometer (Thermo Fisher, USA). The instrument was equipped with spectral acquisition software. In this case, the spectra were acquired in the transmission mode with a liquid cell of a 1-mm light path. Each spectrum consists of an average of 64 scans at 8 cm^{-1} intervals in the wavenumber range of 10,000–4,000 cm^{-1}. We measured NIR spectra of 100 samples: 50 fermented vinegar samples and 50 blended vinegar samples. All the samples were taken directly from the local market in Changsha, P.R. China. The top panel in Figure 5.2 shows the original NIR spectra of all 100 samples, whereas the mean-centered spectra are shown in the bottom panel. Before modeling, the spectra are mean-centered.

However, it is very difficult to detect the difference between the two kinds of samples by visual analysis with human eyes. So, PCA is performed to facilitate a general overview of the clustering of all the samples. PCA is carried out on the mean-centered NIR spectra. The resulting scores plot is shown in Figure 5.3. The star points denote the fermented vinegar samples, and the diamond points stand for the blended samples. It can be seen that although most of the blended samples are well separated from the fermented ones, there are still some overlapping points. The plot is a reflection of the natural structure of the data. To maximize the separation between fermented and blended vinegar samples, supervised pattern recognition should be used. Here, support vector classification is considered.

As discussed in Chapter 2, the choice of a suitable kernel function plays an important role in the performance of SVMs. In this case, we choose the most commonly used radial basis function (RBF) as kernel. So, there are two tuning parameters needing optimization. The first is the penalizing factor C which controls the trade-off between the training

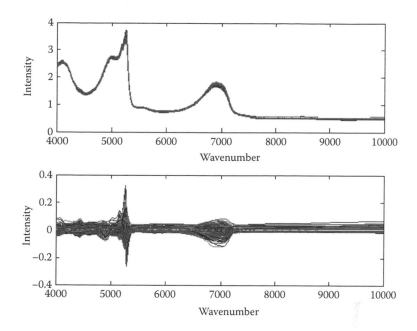

Figure 5.2 The top panel shows the original NIR spectra of the 100 vinegar samples. The centered spectra are shown in the bottom panel.

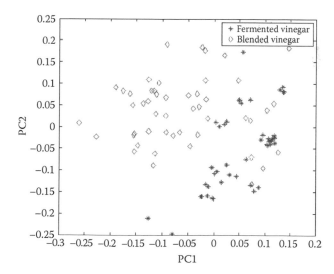

Figure 5.3 The projected samples by principal component analysis.

error and the margin. The second one, γ, is the width associated with the RBF kernel function.

First, 70 samples (35 fermented and 35 blended) are selected as the training set by means of the Kennard–Stone method. The remaining 30 are used to construct a test set. In order to build a SVC model with good performance, the two tuning parameters are first optimized by grid search. The objective is to minimize fivefold cross-validation errors only from the training data. The results are shown in Figure 5.4. In this way, the optimal values of C and γ are 500 and 0.5, respectively. Then we build a SVC model on the training set using the optimized parameters. Both the fitting and test errors are presented in Table 5.1. From the results, it can be found that SVC achieves the minimal cross-validated error and the training error. The prediction errors for the test set of the three methods are the same. On the whole, SVC, for this case, performs best, which may indicate the potential of SVC to deal with the nonlinearity of the real data.

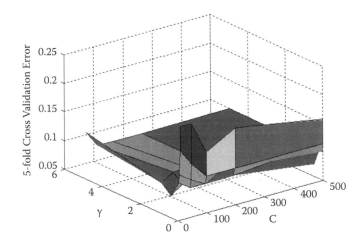

Figure 5.4 The result of grid search for optimizing C and γ.

Table 5.1 Results of PCA-LDA, PLS-LDA, and SVC on the Vinegar Data Using All 1,557 Variables

Methods	CV Error	Training Set	Test Set
PCA-LDA	0.100	0.057	0.033
PLS-LDA	0.086	0.057	0.033
SVC	0.057	0.000	0.033

Note: The numbers of latent variables used in PCA-LDA and PLS-LDA are 6 and 4, respectively. The optimized values of C and γ are 500 and 0.5, respectively.

By the way, the popular way to evaluate the performance of a model is based on an independent test set in the community of statistics and chemometrics. However, in our opinion, it might not be completely correct. The reason is that the prediction error based on a single splitting of the data is severely dependent on both the training and test set. Different partitions of the data will definitely result in different prediction errors. For this reason, we argue that the prediction error computed on a single test is not at all reliable for assessing the predictive performance of a model. One remedy for this problem is to randomly split the data N times, leading to N prediction errors. Then we compare the distribution of the N prediction errors resulting from different methods. In this way, the prediction ability of different models could be reliably assessed. We recently proposed a new concept, called model population analysis (MPA) [33,34], based on which the performance of a model could be comprehensively assessed.

As discussed in Chapter 2, variable selection has proved to be effective for improving the predictive ability of SVMs [35–37]. In this study, the recently proposed strategy, competitive adaptive reweighted sampling (CARS), is first employed to perform variable wavelength selection coupled with PLS [33,38,39]. The source code of CARS-PLS in both MATLAB® and R is freely available at: http://code.google.com/p/carspls/. The output of CARS-PLS is shown in Figure 5.5. The top, middle, and bottom panels show the number of selected variables, the ten fold cross-validated error, and model coefficient path as a function of Monte Carlo sampling runs, respectively. The finally selected variable subset by CARS-PLS contains 11 wavelengths.

To examine whether the selected variables are more discriminating than full spectra, we first perform PCA on the mean-centered reduced data. The scores plot is shown in Figure 5.6. Compared to Figure 5.3, the separation of the two classes of samples based on the 11 variables is much improved, suggesting that variable selection can help identify informative variables and hence benefit classification performance. Interestingly, it can be observed that the first principal component in this case contains very little information with respect to the classification, although it occupies most of the variance (58%). On the contrary, the second principal component, explaining 37% of the total variance, seems to be more informative. The result indicates that the principal component with larger variance is not necessarily the informative one. Furthermore, it also suggests the deficiency of PCA to uncover the useful variation of data responsible for classification.

Besides, PCA-DA, PLS-LDA, and SVC are also applied to the reduced data. The number of latent variables of PCA-DA and PLS-LDA is chosen by means of fivefold cross-validation. For SVC, we also use RBF as the kernel and the two tuning parameters, say, C and γ, are optimized by grid search, which is shown in Figure 5.7. The results are shown in Table 5.2.

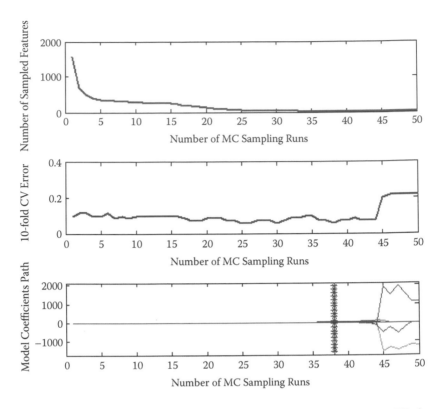

Figure 5.5 The variable selection result of CARS-PLS on the vinegar NIR data. The number of Monte Carlo sampling runs of CARS is set to 50. Top panel: number of variables as a function of iterations. Middle panel: ten fold cross-validated errors at each MC sampling run. Bottom panel: coefficient path of each variable. Each line denotes the path of a variable.

On the whole, the results after variable selection are improved compared to those using the full spectrum. What is important to note here is that, after variable selection, the prediction error on the test set is smaller than the training error. However, in practice, the prediction error should ideally be larger than the training error in that the information of test samples has not been used in the training procedure of the classification model. In our opinion, the results again illustrate that model assessment based on a single splitting of the data is not reliable. We also recommend conducting a model comparison by incorporating the idea of model population analysis by which the performance parameters, for example, prediction error, of different models can be statistically assessed.

Summing up, we have investigated the performance of SVMs in distinguishing blended vinegar samples from fermented samples based on near-infrared spectroscopy. To improve the generalization performance

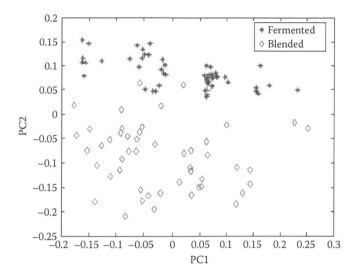

Figure 5.6 Scores plot of all 100 samples based on the 11 variables selected by using CARS-PLS.

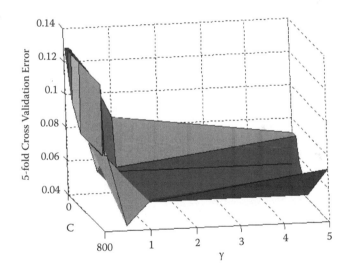

Figure 5.7 Result of grid search for optimizing C and γ using the reduced NIR data of 11 variables.

of SVC, its tuning parameters were optimized by using the grid search method. We also show that the prediction ability of support vector machines could be further enhanced by variable selection. In addition, the performances of two linear discrimination methods, that is, PCA-DA and PLS-LDA, are also examined and compared.

Table 5.2 Results of PCA-LDA, PLS-LDA, and SVC on the
Vinegar Data Using the 11 Variables Selected by CARS-PLSLDA

Methods	CV Error	Training Set	Test Set
PCA-LDA	0.071	0.07	0.033
PLS-LDA	0.057	0.057	0.000
SVC	0.043	0.043	0.000

Note: The numbers of latent variables used in PCA-LDA and PLS-LDA
are 2 and 1, respectively. The optimized values of C and γ are 100
and 1, respectively.

5.3.2   Related work on support vector classification on NIR

In this section, we briefly introduce three classification studies using support vector machines. The first is on both parameter optimization and interpreting the SVM classifiers based on visualization of support vectors (SVs). The second is focused on modeling near-infrared imaging spectroscopy data by means of partial least squares, artificial neural networks, and support vector machines. In the third study, the classification of tomatoes of different genotypes based on SVMs is investigated.

First of all, we give a brief review of the work of Olivier Devos et al. [40]. In their work, the classification performance of SVMs is explored with applications to two NIR datasets. The first one is a slightly nonlinear two-class problem and the second one a more complex three-class task. To guide the choice of parameters of the SVMs, they proposed a methodological approach based on a grid search for minimizing the classification error rate but also relying on the visualization of the number of support vectors. What is worth noting is that they demonstrated the interest of visualizing the SVs in principal component space to go deeper into the interpretation of the trained SVM classifiers. Using the proposed method, the optimized SVM classifiers are very parsimonious and result in good generalization performance on the new samples.

Secondly, an example on the application of SVM in the field of near-infrared imaging spectroscopy (NIRIS) is given here [41]. NIRIS combines the advantages of spectroscopic and microscopic methods along with much faster sample analysis. An imaging spectrometer can at the same time gather both spectral and spatial data by recording sequential images of a given sample. In their study [41], reflectance images were collected in the window of 900–1,700 nm. Pierna et al. compared the performance of support vector machines to those of two other classical classification methods, that is, partial least squares and artificial neural networks (ANNs) in classifying feed particles as either meat and bone meal (MBM) or vegetal using the spectra from NIR images. Although all three methods show good performance in modeling the data, SVMs were found to perform obviously better than PLS and ANN. Their study showed that SVMs were really a good alternative for dealing with the NIR imaging data.

Finally, the application of least squares support vector machines to classifying tomatoes with different genotypes based on visible and near-infrared spectroscopy was reviewed [2]. In Xie, Ying, and Ying's study, visible and short-wave near-infrared (Vis/SW NIR) diffuse reflectance spectroscopic characteristics of tomatoes with different genotypes were measured first. Then, the discrimination performances of different chemometric techniques, including least squares-support vector machines (LS-SVM), discriminant analysis (DA), and soft independent modeling of class analogy (SIMCA) are examined. Their results show that LS-SVM has the same performance as DA. However, the required time of LS-SVM is less if a large variety of tomatoes is to be discriminated.

5.4   Support vector machines for quantitative analysis of near-infrared data

In this section, the quantitative analysis of a benchmark NIR dataset is performed by using support vector regression (SVR) machines. Also, the performance of SVR is compared to that of partial least squares,. currently the most frequently used regression method in chemometrics, especially in the field of NIR analysis. In addition, the performance of SVC and PLS after variable wavelength selection is also presented. Following this, some related work from other groups is also presented.

5.4.1   Correlating diesel boiling points with NIR spectra using SVR

This dataset [42] collected near-infrared spectra of 246 diesel samples with 20 outliers in it. The response value is the boiling point at 50% recovery (BP_{50}). These data were originally obtained at the Southwest Research Institute (SWRI) on a project sponsored by the U.S. Army. In this case, only the 226 normal samples were used for constructing the SVR model. The first-order differential spectra are shown in Figure 5.8. The 226 samples were randomly divided into a training set and a corresponding test set for the construction and evaluation of the SVR model.

PLS is a basic and powerful tool for modeling the linear relationship between the digitalized spectra X and the interesting chemical index y in chemometrics. It can also resist nonlinearity to some extent. In this study, PLS is used as the reference method and the comparative analysis between PLS and SVR is conducted to give some insight into these two methods. The X and y are all scaled into the region [0, 1] before building the PLS and SVR model.

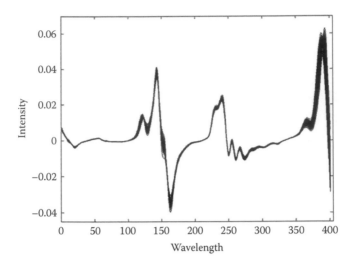

Figure 5.8 The first-order differential NIR spectra of 226 diesel samples.

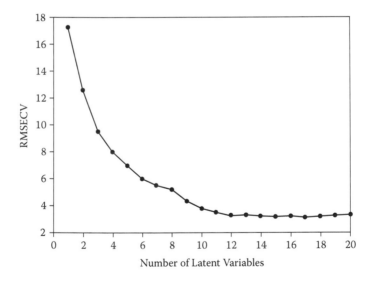

Figure 5.9 RMSECV values of ten fold cross-validation with increasing number of latent variables.

In PLS regression, ten fold cross-validation is employed to choose the number of latent variables (LVs). The root mean squared error of cross-validation (RMSECV) value against the number of LVs is presented in Figure 5.9. From this plot, it can easily be found that the RMSECV value begins to increase after 12 LVs. Therefore, 12 LVs are

Table 5.3 Results of SVR for NIR Data[a]

	Train	Test	Train[b]	R^2
PLS	2.3397	2.9431	3.2571	0.9757
SVMs	1.8209	2.6621	3.2417	0.9835

[a] The optimized parameters of SVR for NIR data are: C = 59.69, v = 0.3292. Notice that the original *y*-values are scaled into the region [0, 1] before building the PLS and SVR model. The results in this table are reported after transforming the scaled *y*-value into the original unit.

[b] RMSECV values based on ten fold cross-validation.

chosen to construct the PLS model using the training set. Then an independent test set is used to evaluate the prediction ability of the established model.

v-SVR is applied here to construct the SVR model. The RBF is used as the kernel. The γ of the RBF kernel is set to the default value in the LIBSVM software [43]. There are two parameters to be predefined before training. One is the regularizing factor C; the other is the sparsity parameter v. The genetic algorithm (GA) is used in this study to optimize these two parameters globally. The optimized values of C and v by GA are 59.69 and 0.3292, respectively (Table 5.3). Finally, the SVR model is built with the optimized parameters using the training set. Results obtained by both PLS and SVR are shown in Table 5.3. The fitting and prediction errors of SVR for the training set and test set are lowered by 22.1% and 9.5% compared to those of PLS, which indicates an obvious improvement. The squared correlative coefficients R^2 of SVR and PLS on the whole dataset are 0.9835 and 0.9737, respectively. This result further reveals the better prediction ability of SVR.

The predicted and experimental values are shown in Figure 5.10. From this figure, it can be seen that SVR gives both better fitting and prediction results. It may be concluded that SVR not only has the ability to model the main linear relationship between **X** and **y** but also can grasp the nonlinearity existing in real-world data. But PLS can't do this well in that PLS is inherently a linear modeling method, which makes it impossible to account well for the nonlinear part of the data. Despite the good prediction performance, it should be pointed out that SVR has its limitations and disadvantages when dealing with a spectral dataset. First, SVR is nonlinear, and this makes it difficult for the researcher to explain the results, for example, which region or combination of different regions is really meaningful for the model. Second, the optimization of parameters of SVR is a relatively time-consuming task compared to PLS, which may limit their applications. Conclusively, SVR can be seen as a competitive and promising method for modeling NIR data of some nonlinearity.

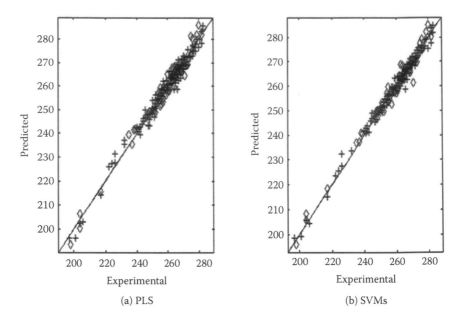

Figure 5.10 Predicted values versus experimental values for diesel boiling point data: (a) PLS; (b) SVMs.

5.4.2   Related work on support vector regression on NIR

There is increasing evidence showing that near-infrared spectroscopy has been gaining wide application in a number of fields. We present here some work on modeling near-infrared data for practical purposes by using support vector machines.

In the work of R. J. Poppi et al. [8], least squares-support vector machines are proposed as an alternative multivariate calibration method for the simultaneous quantification of some common adulterants (starch, whey, or sucrose) found in powdered milk samples. The powdered milk samples used in their study were supplied by DISCAMP-Brazil and contaminated in bulk with milk whey, starch, or sucrose at different concentration levels. After the samples were prepared, the diffuse reflectance spectra were recorded on a CARY 5G UV/VIS/NIR spectrophotometer, set in the NIR region (1027.5–2400.0 nm), using Teflon as reference. For each sample, three replicate spectra were obtained. As the authors stated, because the spectral differences of the three adulterants took on a non-linear behavior when all groups of adulterants were in the same dataset, the use of linear methods such as partial least squares regression (PLSR) was challenged. In this case, nonlinear support vector machines may be a better choice. By comparative study, they found that LS-SVM resulted in excellent models, with low prediction errors and superior performance to PLSR. Their results demonstrated that it was possible to build robust models for quantitative analysis of some common adulterants in powdered milk by means of a combination of near-infrared spectroscopy and LS-SVM [8].

The second example is chosen from the work of Perez-Marin et al. on NIR spectroscopy calibration in the feed industry [44]. In this area, the development of NIRS calibrations to predict the ingredient composition of compound feeds is a complex task, because of both the complicated nature of the parameters to be predicted and the heterogeneous nature of the matrices/formulas. Thus, multivariate linear calibration methods do not work well. To develop NIRS models for predicting two of the most representative ingredients (i.e., wheat and sunflower meal), the authors evaluated the performance of three nonlinear methods: least squares-support vector machines, CARNAC, and locally biased regression. Their results showed that the best performance is achieved by CARNAC. LS-SVM performed less well than CARNAC. And locally biased regression was the least successful among the three methods. When compared to the standard approach by using PLS, the prediction results of all three methods, however, are improved, suggesting that the choice of nonlinear methods for modeling complex data is necessary.

5.5   Some comments

Although SVMs have found successful application in many fields, it should be admitted, however, that there is still a long way for SVMs to go before large-scale applications are feasible. The reasons lie in several aspects, including but not limited to (1) the procedure for optimizing tuning parameters of SVMs seems a little bit complex and time-consuming, which may be a major factor limiting the use of SVM; (2) the SVM classification or regression model using SVMs is hard to interpret because of its nonlinear nature (nonlinear kernel); (3) the classification boundaries for SVMs are often highly unstable and heavily dependent on the samples included in the training set, with the result that their inclusion or exclusion in the training set could make a big difference to the model and therefore to the classification accuracy. Regarding these problems, it's recommended that the practitioners of SVMs should at least have at least a basic understanding of the mechanism before utilizing SVMs for any analysis.

With respect to the assessment of the SVM model, one widely accepted way is to compute the prediction error on an independent test set. However, in our opinion, the assessment of predictive performance based on a single test set is not convincing and rather dangerous unless the size of both the training and test sets is large enough (this

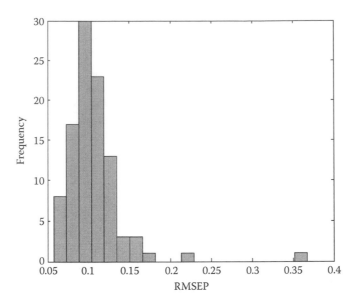

Figure 5.11 The distribution of 100 RMSEP values resulting from random splitting.

is usually impossible in practice). To illustrate this point, benchmark NIR data are used here to show the high variability of prediction error resulting from different training/test set pairs. We used the corn data that can be freely downloaded from www.eigenvector.com. The chosen data contained NIR spectra of 80 corn samples measured on an mp5 instrument with the target of modeling protein content. We randomly split the data into calibration samples (64 samples) and a corresponding test set (16 samples). We chose nu-SVR to build SVM models. The RBF was chosen as the kernel. The training set was used to optimize the tuning parameters using the grid search method and then we built a calibration model, which was used to compute the prediction error on the test set. The root mean squared error of prediction (RMSEP) was recorded. We repeated this procedure 100 times, and 100 RMSEP are values, the distribution of which is shown in Figure 5.11. RMSEP values varied greatly from 0.05 to 0.35, suggesting the use of a single test set to estimate prediction error is rather dangerous and hence not convincing, especially for nonlinear modeling methods such as support vector machines. For this reason, it's not recommended to conduct model assessment or comparison based on only a single test. Much more work is needed in the field of model assessment.

References

1. Donald, A. B. and Emil, W. C. 2008. *Handbook of Near-Infrared Analysis*, 3rd edn. Boca Raton, FL: CRC Press.
2. Xie, L.J., Ying, Y.B., and Ying, T.J. 2009. Classification of tomatoes with different genotypes by visible and short-wave near-infrared spectroscopy with least-squares support vector machines and other chemometrics. *J. Food Eng.* 94(1):34–39.
3. Paris, I., Janoly-Dumenil, A., Paci, A., Mercier, L., Bourget, P., Brion, F., Chaminade, P., and Rieutord, A. 2006. Near infrared spectroscopy and process analytical technology to master the process of busulfan paediatric capsules in a university hospital. *J. Pharmaceut. Biomed.*, 41(4):1171–1178.
4. Luo, X., Yu, X., Wu, X., Cheng, Y., and Qu, H. 2008. Rapid determination of *Paeoniae radix* using near infrared spectroscopy. *Microchem. J.*, 90(1):8–12.
5. Reich, G. 2005. Near-infrared spectroscopy and imaging: Basic principles and pharmaceutical applications. *Adv. Drug. Deliver. Rev.*, 57(8):1109–1143.
6. Amerov, A.K., Chen, J., Small, G.W., and Arnold, M.A. 2005. Scattering and absorption effects in the determination of glucose in whole blood by near-infrared spectroscopy. *Anal. Chem.*, 77(14):4587–4594.
7. Alessandrini, L., Romani, S., Pinnavaia, G., and Rosa, M.D. 2008. Near infrared spectroscopy: An analytical tool to predict coffee roasting degree. *Anal. Chim. Acta*, 625(1):95–102.
8. Borin, A., Ferrão, M.F., Mello, C., Maretto, D.A., and Poppi, R.J. 2006. Least-squares support vector machines and near infrared spectroscopy for quantification of common adulterants in powdered milk. *Anal. Chim. Acta*, 579(1):25–32.
9. Khodabux, K., L'Omelette, M.S.S., Jhaumeer-Laulloo, S., Ramasami, P., and Rondeau, P. 2007. Chemical and near-infrared determination of moisture, fat and protein in tuna fishes. *Food Chem.*, 102(3):669–675.
10. Wu, D., Cao, F., Feng, S.-J., and He, Y. 2008. Application of infrared spectroscopy technique to protein content fast measurement in milk powder based on support vector machines. *Guang Pu Xue Yu Guang Pu Fen Xi*, 28(5):1071–1075.
11. Savenije, B., Geesink, G.H., van der Palen, J.G.P., and Hemke, G. 2006. Prediction of pork quality using visible/near-infrared reflectance spectroscopy. *Meat Sci.*, 73(1):181–184.
12. Watson, C.A. 1977. Near infrared reflectance spectrophotometric analysis of agricultural products. *Anal. Chem.*, 49(9):835A–840A.
13. Bellon, V., Vigneau, J.L., and Sila, F. 1994. Infrared and near-infrared technology for the food industry and agricultural uses: On-line applications. *Food Control*, 5(1):21–27.
14. Reeves, J.B., McCarty, G.W., and Meisinger, J.J. 1999. Near infrared reflectance spectroscopy for the analysis of agricultural soils. *J. Near Infread Spec.*, 7(3):179–193.
15. Walsh, K.B., Golic, M., and Greensill, C.V. 2004. Sorting of fruit using near infrared spectroscopy: Application to a range of fruit and vegetables for soluble solids and dry matter content. *J. Near Infread Spec.*, 12(3):141–148.
16. Frank, I.E. and Friedman, J.H. 1993. A statistical view of some chemometrics regression tools. *Technometrics*, 35:109–148.

17. Hartnett, M.K., Lightbody, G., and Irwin, G.W. 1998. Dynamic inferential estimation using principal components regression (PCR). *Chemometr. Intell. Lab.*, 40(2):215–224.
18. Gemperline, P.J. and Salt, A. 1989. Principal components regression for routine multicomponent UV determinations: A validation protocol. *J. Chemometr.*, 3(2):343–357.
19. Xu, Q.-S., Liang, Y.-Z.L., and Shen, H.-L. 2001. Generalized PLS regression. *J. Chemometr.*, 15(3):135–148.
20. Geladi, P. and Kowalski, B.R. 1986. Partial least-squares regression: A tutorial. *Anal. Chim. Acta*, 185:1–17.
21. Wold, S., Sjöström, M., and Eriksson, L. 2001. PLS-regression: A basic tool of chemometrics. *Chemometr. Intell. Lab.*, 58(2):109–130.
22. De Jong, S. 1993. SIMPLS: An alternative approach to partial least squares regression. *Chemometr. Intell. Lab.*, 18(3):251–263.
23. Khan, J., Wei, J.S., Ringner, M., Saal, L.H., Ladanyi, M., Westermann, F., Berthold, F., Schwab, M., Antonescu, C.R., Peterson, C., et al. 2001. Classification and diagnostic prediction of cancers using gene expression profiling and artificial neural networks. *Nature Med.*, 7(6):673–679.
24. Yao, X. 1999. Evolving artificial neural networks. *Proc. IEEE*, 87(9):1423–1447.
25. Hopfield, J.J. 1982. Neural networks and physical systems with emergent collective computational abilities. *Proc. Nat. Acad. Sci. USA Biol. Sci.*, 79(8):2554–2558.
26. Vapnik, V. 1999. *The Nature of Statistical Learning Theory*, second edition. New York: Springer.
27. Vapnik, V. 1998. *Statistical Learning Theory*. New York: Wiley.
28. Li, H.-D., Liang, Y.-Z., and Xu, Q.-S. 2009. Support vector machines and its applications in chemistry. *Chemometr. Intell. Lab.*, 95:188–198.
29. Xu, Y., Zomer, S., and Brereton, R. 2006. Support vector machines: A recent method for classification in chemometrics. *Crit. Rev. Anal. Chem.*, 36:177–188.
30. Burges, C. 1998. A tutorial on support vector machines for pattern recognition. *Data Min. Knowl. Disc.*, 2:121–167.
31. Smola, A.J. and Scholkopf, B. 2004. A tutorial on support vector regression. *Statist. Comput.*, 14:199–222.
32. Hasegawa, K. and Funatsu, K. Non-linear modeling and chemical interpretation with aid of support vector machine and regression. *Curr. Comput. Aid. Drug Des.*, 6(1):24–36.
33. Li, H.-D., Liang, Y.-Z., Xu, Q.-S., Cao, D.-S. 2010. Model population analysis for variable selection. *J. Chemometr.*, 24(7–8):418–423.
34. Li, H.-D., Zeng, M.-M., Tan, B.-B., Liang, Y.-Z., Xu, Q.-S., and Cao, D.-S. 2010. Recipe for revealing informative metabolites based on model population analysis. *Metabolomics*, 6(3) 353–361.
35. Guyon, I., Weston, J., Barnhill, S., and Vapnik, V. 2002. Gene selection for cancer classification using support vector machines. *Mach. Learn.*, 46(1):389–422.
36. Bierman, S. and Steel, S. 2009. Variable selection for support vector machines. *Commun. Statist. Simul. Comput.*, 38(8):1640–1658.
37. Zhang, H.H. 2006. Variable selection for support vector machines via smoothing spline ANOVA. *Statist. Sinica*, 16(2):659–674.

38. Li, H.-D., Liang, Y.-Z., Xu, Q.-S., and Cao, D.-S. 2009. Key wavelengths screening using competitive adaptive reweighted sampling method for multivariate calibration. *Anal. Chim. Acta*, 648(1):77–84.
39. http://code.google.com/p/carspls/.
40. Devos, O., Ruckebusch, C., Durand, A., Duponchel, L., and Huvenne, J.P. 2009. Support vector machines (SVM) in near infrared (NIR) spectroscopy: Focus on parameters optimization and model interpretation. *Chemometr. Intell. Lab. Syst.*, 96(1):27–33.
41. Pierna, J.A.F., Baeten, V., Renier, A.M., Cogdill, R.P., and Dardenne, P. 2004. Combination of support vector machines (SVM) and near-infrared (NIR) imaging spectroscopy for the detection of meat and bone meal (MBM) in compound feeds. *J.Chemometr.*, 18(7–8):341–349.
42. http://software.eigenvector.com/Data/index.html.
43. http://www.csie.ntu.edu.tw/~cjlin/libsvm.
44. Perez-Marin D., Garrido-Varo, A., Guerrero, J. E., Fearn, T. , and Davies, A. M. C. 2008. Advanced nonlinear approaches for predicting the ingredient composition in compound feedingstuffs by near-infrared reflection spectroscopy. *Applied Spectroscopy*, 62(5)536–541.

chapter six

Support vector machines and QSAR/QSPR

Contents

6.1 Introduction

In this chapter, we begin by introducing the history of quantitative structure-activity/property relationship (QSAR/QSPR) and molecular descriptors. Then the principles for QSAR modeling are described in some detail to facilitate the consideration of the QSAR/QSPR model for regulatory purposes. The related QSAR/QSPR studies using SVMs are also reviewed. Finally, two real QSAR datasets for regression and classification are employed to illustrate the outstanding performance of SVMs as a useful QSAR modeling tool.

6.2　Quantitative structure-activity/property relationship

6.2.1　History of QSAR/QSPR and molecular descriptors

The history of QSAR/QSPR and molecular descriptors is closely related to the history of what can be considered one of the most important scientific concepts of the last part of the nineteenth century and the whole of the twentieth century, that is, the concept of molecular structure. The years between 1860 and 1880 were characterized by a strong dispute about the concept of molecular structure, arising from the studies on substances showing optical isomerism and the studies of Kekulé (1861–1867) on the structure of benzene. The concept of the molecule as a three-dimensional (3-D) body was first proposed by Butlerov (1861–1865), Wislicenus (1869–1873), Van't Hoff (1874–1875), and Le Bel (1874). The publication in French of the revised edition of *La chimie dans l 'éspace* by Van't Hoff in 1875 is considered a milestone of the 3-D concept of chemical structures.

QSAR history started a century earlier than the history of molecular descriptors, being closely related to the development of the molecular structure theories. QSAR modeling was born in the field of toxicology. Attempts to quantify relationships between chemical structure and acute toxic potency have been part of the toxicological literature for more than 100 years. In the defense of his thesis entitled, "Action de l'alcohol amylique surl'organisme," at the Faculty of Medicine, University of Strasbourg, France, on January 9, 1863, Cros noted that a relationship existed between the toxicity of primary aliphatic alcohols and their water solubility. This relationship demonstrated the central axiom of structure–toxicity modeling; that is, the toxicity of substances is governed by their properties, which are determined in turn by their chemical structure. Therefore, there are interrelationships among structure, properties, and toxicity.

Crum-Brown and Fraser (1868–1869) [1,2] proposed the existence of a correlation between biological activity of different alkaloids and their molecular constitution. More specifically, the physiological action of a substance in a certain biological system (Φ) was defined as a function (f) of its chemical constitution (C):

$$\Phi = f(C)$$

Thus, an alteration in chemical constitution ΔC would be reflected by an effect on biological activity $\Delta\Phi$. This equation can be considered the first general formulation of a QSAR.

A few years later, an hypothesis on the existence of correlations between molecular structure and physicochemical properties was reported in the work of Körner (1874) [3], which dealt with the synthesis of disubstituted benzenes and the discovery of ortho-, meta-, and para-derivatives. The different colors of disubstituted benzenes were thought to be related to their differences in molecular structure. Ten years later, Mills (1884) [4] published a study, "On melting point and boiling point as related to composition," in the *Philosophical Magazine*. The quantitative property–activity models, commonly considered to mark the beginning of systematic QSAR/QSPR studies [5], have emerged from the search for relationships between the potency of local anesthetics and the oil/water partition coefficient between narcosis and chain length, and narcosis and surface tension. In particular, the concepts developed by Meyer and Overton are often referred to as the Meyer–Overton theory of narcotic action.

The first theoretical QSAR/QSPR approaches date back to the end of the 1940s and relate biological activities and physicochemical properties to theoretical numerical indices derived from the molecular structure. The first theoretical molecular descriptors were based on graph theory: the Wiener index [6] and the Platt number [7] proposed in 1947 to model the boiling point of hydrocarbons. In the early 1960s, new molecular descriptors were proposed, giving the start to systematic studies on the molecular descriptors, mainly based on graph theory. The use of quantum chemical descriptors in QSAR studies dates back to the early 1970s [8], although quantum chemical descriptors were defined and used a long time before in the framework of quantum chemistry. During 1930–1960, the milestones were the works of Pauling [9] and Coulson on the chemical bond, Sanderson [10] on electronegativity, and Fukui, Yonezawa, and Shingu [11] and Mulliken [12] on electronic distribution. Once the concept of molecular structure was definitively consolidated by the successes of quantum chemistry theories and the approaches to the calculation of numerical indices encoding molecular structure information were accepted, all the constitutive elements for the take-off of QSAR strategies were available. Based on the Hammett equation [13], the seminal work of Hammett gave rise to the "$\sigma - \rho$" culture in the delineation of substituent effects on organic reactions, whose aim was the search for linear free-energy relationships (LFERs) [14]: steric, electronic, and hydrophobic constants were defined, becoming basic tools for modeling properties of molecules.

In the 1950s, the fundamental works of Taft [15,16] in physical organic chemistry were the foundation of relationships between physicochemical properties and solute–solvent interaction energies (linear solvation energy relationships, LSERs), based on steric, polar, and resonance parameters for substituent groups in congeneric compounds. In the mid-1960s, led by the pioneering works of Hansch et al. [17,18], the QSAR/QSPR approach began to assume its modern look. In 1962, Hansch and coworkers [17] published their study on the structure–activity relationships of plant growth regulators and their dependency on Hammett constants and hydrophobicity. Using the octanol/water system, a whole series of partition coefficients was measured, and thus, a new hydrophobic scale was introduced for describing the aptitude of molecules to move through environments characterized by different degrees of hydrophilicity such as blood and cellular membranes. The delineation of Hansch models led to explosive development in QSAR analysis and related approaches.

At the same time, Free and Wilson [19] developed a model of additive substituent contributions to biological activities, giving a further push to the development of QSAR strategies. They proposed to model a biological response on the basis of the presence or absence of substituent groups on a common molecular skeleton. This approach, called the "de novo approach" when presented in 1964, was based on the assumption that each substituent gives an additive and constant effect to the biological activity regardless of the other substituents in the rest of the molecule. By the end of the 1960s, many structure–property relationships were proposed based not only on substituent effects but also on indices describing the whole molecular structure. These theoretical indices were derived from a topological representation of the molecule, mainly applying graph theory concepts, and then usually referred to as 2-D descriptors.

The fundamental works of Balaban and Harary [20], Balaban [21], Randic [22,23], and Kier et al. [24] led to further significant developments of the QSAR approaches based on topological indices (TIs). As a natural extension of the topological representation of a molecule, the geometrical aspects have been taken into account since the mid-1980s, leading to the development of the 3-D-QSAR, which exploits information on molecular geometry. Geometrical descriptors were derived from the 3-D spatial coordinates of a molecule, and among them, there were shadow indices [25], charged partial surface area descriptors [26], weighted holistic invariant molecular (WHIM) descriptors [27], eigenvalue (EVA) descriptors [28], 3-D-MoRSE descriptors [29], and geometry, topology, and atom-weight assembly (GETAWAY) descriptors [30].

In the late 1980s, a new strategy for describing molecule characteristics was proposed, based on molecular interaction fields (MIFs), which are composed of interaction energies between a molecule and probes, at specified spatial points in 3-D space. Different probes (such as a water molecule, methyl group, and hydrogen) were used for evaluating the interaction energies in thousands of grid points where the molecule was embedded. As a final result of this approach, a scalar field (a lattice) of interaction energy values characterizing the molecule was obtained. The first formulation of a lattice model to compare molecules by aligning them in 3-D space and extracting chemical information from MIF was proposed by Goodford [31] in the GRID method and then by Cramer, Patterson, and Bunce [32] in the comparative molecular field analysis (CoMFA). Still based on MIFs, several other methods were successively proposed, among them comparative molecular similarity indices analysis (CoMSIA) [33], the Compass method [34], GRID descriptors [35], and so on. Finally, increasing interest on the part of the scientific community has been shown in recent years for combinatorial chemistry, high-throughput screening, substructural analysis, and similarity searching, for which several similarity/diversity approaches have been proposed mainly based on substructure descriptors such as molecular fingerprints [36,37].

6.2.2   Principles for QSAR modeling

In November 2004, the 37th OECD's Joint Meeting of the Chemicals Committee and the Working Party on Chemicals, Pesticides and Biotechnology (Joint Meeting) agreed on the *OECD Principles for the Validation, for Regulatory Purposes, of (Q)SAR Models*. Flexibility will be needed in the interpretation and application of each OECD principle because, ultimately, the proper integration of (Q)SARs into any type of regulatory or decision-making framework depends upon the needs and constraints of the specific regulatory authority. For example, the need for such flexibility is given in a case study by the U.S. EPA on the regulatory uses and applications of (Q)SAR models in OECD member countries (OECD, 2006). The document can be freely downloaded at the website (http://www.oecd.org/searchResult/0,3400,en_2649_201185_1_1_1_1_1,00.html).

These agreed OECD principles are as follows.

> To facilitate the consideration of a (Q)SAR model for regulatory purposes, it should be associated with the following information:
>
> 1. a defined endpoint;
> 2. an unambiguous algorithm;
> 3. a defined domain of applicability;

4. appropriate measures of goodness-of-fit, robustness and predictivity;
5. a mechanistic interpretation, if possible.

According to Principle 1, a (Q)SAR should be associated with a defined endpoint, where endpoint refers to any physicochemical, biological, or environmental effect that can be measured and therefore modeled. The intent of this principle is to ensure transparency in the endpoint being predicted by a given model, inasmuch as a given endpoint could be determined by different experimental protocols and under different experimental conditions. Ideally, (Q)SARs should be developed from homogeneous datasets in which the experimental data have been generated by a single protocol. However, this is rarely feasible in practice, and data produced by different protocols are often combined.

According to Principle 2, a (Q)SAR should be expressed in the form of an unambiguous algorithm. The intent of this principle is to ensure transparency in the description of the model algorithm. In the case of commercially developed models, this information is not always made publicly available.

According to Principle 3, a (Q)SAR should be associated with a defined domain of applicability. The need to define an applicability domain expresses the fact that (Q)SARs are reductionist models that are inevitably associated with limitations in terms of the types of chemical structures, physicochemical properties, and mechanisms of action for which the models can generate reliable predictions. This principle does not imply that a given model should only be associated with a single applicability domain.

According to Principle 4, a (Q)SAR should be associated with appropriate measures of goodness-of-fit, robustness, and predictivity. This principle expresses the need to provide two types of information: (a) the internal performance of a model (as represented by goodness-of-fit and robustness), determined by using a training set; and (b) the predictivity of a model, determined by using an appropriate test set.

According to Principle 5, a (Q)SAR should be associated with a mechanistic interpretation, wherever such an interpretation can be made. Clearly, it is not always possible to provide a mechanistic interpretation of a given (Q)SAR. The intent of this principle is therefore to ensure that there is an assessment of the mechanistic associations between the descriptors used in a model and the endpoint being predicted, and that any association is documented. Where a mechanistic interpretation is possible, it can also form part of the defined applicability domain (Principle 3).

6.3   Related QSAR/QSPR studies using SVMs

SVMs, as one of the most popular machine learning approaches, have been widely applied to the studies of QSAR/QSPR for classification and regression. They have been shown to exhibit low prediction error in QSAR/QSPR [38,39]. Here, we choose to briefly review some related QSAR/QSPR studies by using SVMs. These study works are mainly focused on the prediction of the activities or ADMET properties of drug compounds. Studies of P-glycoprotein substrates used SVMs [40,41] with more accurate results than neural networks, decision trees, and k-NN. A study focused on prediction of drug likeness [42], has shown lower prediction error for SVM than for bagging ensembles and for linear methods. In a study involving COX-2 inhibition and aquatic toxicity [43], SVMs outperformed MLR and RBF neural networks. An extensive study using SVMs among other machine-learning methods was conducted [44,45] using a wide range of endpoints. In this study, SVMs were usually better than k-NN, decision trees, and linear methods, but slightly inferior to boosting and random forest ensembles. SVMs have also been tested with ADME properties, including modeling of human intestinal absorption [41], binding affinities to human serum albumin [46], and cytochrome P450 inhibition [47]. Studies focused on hemostatic factors have employed SVMs, for example, modeling thrombin binding [48] and factor Xa binding [49]. Adverse drug effects, such as carcinogenic potency [50] and Torsade de Pointes [41,46] were analyzed using the SVM method. Properties such as O–H bond dissociation energy in substituted phenols [51] and capillary electrophoresis absolute mobilities [52] have also been studied using SVMs, which exhibited higher accuracy than linear regression and RBF neural networks. Other properties predicted with SVM include heat capacity [53] and capacity factor (log k) of peptides in high-performance liquid chromatography [54].

6.4 Support vector machines for regression

In this section, we mainly analyze a real-world example related to aqueous solubility of druglike molecules by using support vector regression (SVR). During model-building, rigorous steps are performed according to OECD principles for QSAR/QSPR modeling, including data collection, molecule description, descriptor selection, model internal validation, model building, applicability domain, and model interpretation. To demonstrate the prediction performance of SVR, partial least squares (PLS), and back-propagation network (BPN) are also employed to compare the results of SVR.

6.4.1 Dataset description

All druglike molecules used in this study were collected from two sources. The first one reported the aqueous solubility of a diverse set of 88 druglike molecules at 25°C and an ionic strength of 0.15 M using the CheqSol approach which is a highly reproducible potentiometric technique that ensures the thermodynamic equilibrium is reached rapidly. (A total of 101 druglike molecules were downloaded from http://www-jmg.ch.cam. ac.uk/data/. Among them, 9 molecules lacked aqueous solubility and 4 molecules were measured under different experimental conditions. So, these 13 druglike molecules were excluded.) Results with a coefficient of variation higher than 4% were rejected. Moreover, the experimental aqueous solubility (Sol) data measured at 298 K and expressed in mg/ml for 137 structurally diverse druglike organic compounds were extracted from the Merck index [55]. Solubility data were checked at ChemID Plus (National Library of Medicine, National Institutes of Health). No differences in solubility data were found between the Merck index and ChemID records. So, the aqueous solubilities of the total 225 druglike molecules are credible enough and we can use them to establish a reliable model.

6.4.2   Molecular modeling and descriptor calculation

The structures of all molecules were firstly preoptimized with the molecular mechanics force field (MM+) procedure included in the Hyperchem 7.0 package, and the resulting geometries were further refined by means of the semi-empirical Molecular Orbital's Method PM3 (Parametric Method-3) using the Polak–Ribiere algorithm with a gradient norm limit of 0.01 kcal/Å. Finally, 1,640 molecular descriptors were obtained using the software Dragon 5.4 [13], including all types of descriptors such as constitutional, topological, geometrical, informational, charge, GETAWAY, RDF, WHIM, 3-D-MoRSE, walk and path counts, 2-D autocorrelations, Randic molecular profiles, atom-centered fragments, edge adjacency indices, functional group counts, connectivity indices, burden eigenvalues, molecular properties, and so on.

To ensure that a derived model has a good generalization ability, all druglike molecules are split into two parts, namely an internal training set and external validation set (or test set), using the Kennard–Stone (KS) method [56]. The KS method is usually used to select a representative subset from the total datasets due to its good performance in other studies. Finally 180 druglike molecules were used as the training set and remainder used as the test set.

6.4.3   Feature selection using a generalized cross-validation program

A generalized cross-validation feature selection program was used to select the most suitable molecular descriptors that relate to the aqueous solubility of 180 druglike compounds. Finally, 28 molecular descriptors were picked out from the total pool of molecular descriptors. The descriptors calculated by Dragon software are listed in Table 6.1.

Table 6.1 Symbols for 28 Molecular Descriptors Used in the Study

Molecular Descriptor	Type	Description
PCD	Walk and path counts	Difference between multiple path count and path count
ATS1m	2-D autocorrelations	Broto–Moreau autocorrelation of a topological structure – lag 1/weighted by atomic masses
MATS2m	2-D autocorrelations	Moran autocorrelation – lag 2/weighted by atomic masses
MATS7m	2-D autocorrelations	Moran autocorrelation – lag 7/weighted by atomic masses
MATS3e	2-D autocorrelations	Moran autocorrelation – lag 3/ weighted by atomic Sanderson electronegativities
BELe4	Burden eigenvalues	Lowest eigenvalue n. 4 of Burden matrix/ weighted by atomic Sanderson electronegativities
GGI1	Topological charge indices	GGI1 topological charge index of order 1
GGI9	Topological charge indices	Topological charge index of order 9
VRp2	Eigenvalue-based indices	Average Randic-type eigenvector-based index from polarizability weighted distance matrix
RGyr	Geometrical descriptors	Radius of gyration (mass weighted)
RCI	Geometrical descriptors	Jug RC index
RDF105u	RDF descriptors	Radial distribution function – 10.5/ unweighted
RDF020m	RDF descriptors	Radial distribution function – 2.0/ weighted by atomic masses
RDF065m	RDF descriptors	Radial distribution function – 6.5/ weighted by atomic masses
RDF110m	RDF descriptors	Radial distribution function – 11.0/ weighted by atomic masses
RDF015e	RDF descriptors	Radial distribution function – 1.5/ weighted by atomic Sanderson electronegativities
RDF100e	RDF descriptors	Radial distribution function – 10.0/ weighted by atomic Sanderson electronegativities

(Continued)

Table 6.1 Symbols for 28 Molecular Descriptors Used in the Study (Continued)

Molecular Descriptor	Type	Description
Mor20u	3-D-MoRSE descriptors	3D-MoRSE – signal 20/unweighted
Mor27p	3-D-MoRSE descriptors	3D-MoRSE –signal 27/weighted by atomic polarizabilities
HATS8v	GETAWAY descriptors	Leverage-weighted autocorrelation of lag 8/weighted by atomic van der Waals volumes
R3v	GETAWAY descriptors	R autocorrelation of lag 3/weighted by atomic van der Waals volumes
R4e	GETAWAY descriptors	R autocorrelation of lag 4/ weighted by atomic Sanderson electronegativities
C-026	Atom-centered fragments	R—CX—R
C-040	Atom-centered fragments	R–C(=X)–X/R–C#X/X = C = X
H-052	Atom-centered fragments	H attached to C0(sp3) with 1X attached to next C
O-058	Atom-centered fragments	=O
ALOGP	Molecular properties	Ghose–Crippen octanol–water partition coeff. (log P)
ALOGP2	Molecular properties	Squared Ghose–Crippen octanol–water partition coeff. (log P^2)

When the predictor variables are highly correlated in a model, our estimated regression coefficients tend to have large sampling variability. Thus the estimated regression coefficients tend to vary widely from one sample to the next. In addition, the common interpretation of a regression coefficient, as measurement of the change in the expected value of the response variable, is not fully applicable when multicollinearity exists [57]. So the variance inflation factor (VIF) is used to detect the multicollinearity of 28 predictor variables. The largest VIF value among all variables is often used as an indicator of the severity of multicollinearity, A maximum VIF value in excess of 20 is frequently taken as an indication that multicollinearity might unduly influence the least squares estimator. Figure 6.1 shows the variance inflation factor for every descriptor. We can see that VIF for all molecular descriptors varied from 0 to 20, which indicates that

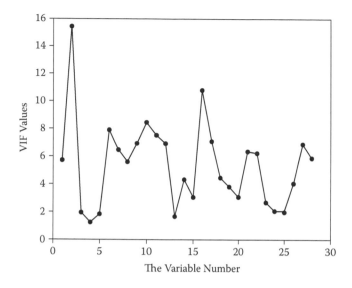

Figure 6.1 The values of VIF for every molecular descriptor used in the study.

multicollinearity of 28 molecular descriptors does not significantly affect the estimation of regression coefficients.

6.4.4 Model internal validation

In order to demonstrate that the established model represented real structure–property relationships, ten fold cross-validation was adopted in this work for model assessment [58]. For ten fold cross-validation, the training set was first split into 10 roughly equal-sized parts and then the model fit to nine parts of the data and the prediction error of the remaining part calculated. The process was repeated 10 times so that every part could be used as a validation set. The parameters employed to evaluate the behavior in this investigation were four commonly used ones in regression problems, including the square correlation coefficients of fitting (R^2), the square correlation coefficients of cross-validation (Q^2), the root-mean-squared error of fitting (RMSEF), and the root-mean-squared error of cross-validation (RMSECV). The latter two are calculated by means of the formulas:

$$\text{RMSEF} = \sqrt{\frac{1}{N} \sum_{i=1}^{N} (y_i - \hat{y})^2}$$

Here N is the number of samples in the training set and $\hat{y}_i$ is the predictive value of the ith sample in the training set.

$$\text{RMSECV} = \sqrt{\frac{1}{N} \sum_{1}^{v} (y_v - \hat{y}_{v(v)})^2}$$

where v is the number of the fold of cross-validation and $\hat{y}_{v(v)}$ is the predictive value of samples of the vth fold when the vth fold samples are left out.

6.4.5   PLS regression model

Partial least squares regression analysis was first performed using 28 descriptors as input variables for 180 druglike compounds. The number of significant latent variables for the PLS algorithm was determined using ten fold cross-validation. The following equation was obtained and the corresponding regression coefficients were plotted in Figure 6.2:

$$\text{Log S} = \sum (c_i D_i) + 8.2611$$

where D_i represents a molecular descriptor and c_i is its corresponding regression coefficient in the PLS model. For the training set, $R^2 = 0.8457$, RMSEF = 0.5993, and for the cross-validation set, $Q^2 = 0.767$, RMSECV = 0.737.

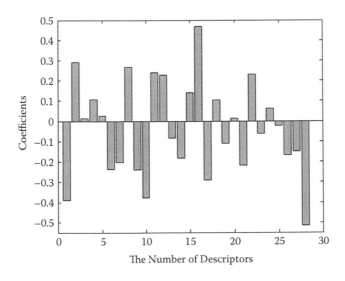

Figure 6.2 The regression coefficient plot of the PLS model.

6.4.6 BPN regression model

A standard back-propagation network was applied to predict the aqueous solubility. An input layer with 28 input units, an output layer with one neuron representing log S, and a hidden layer of several neurons were used. All layers were completely connected. The initial weights were randomly initialized between –0.5 and 0.5. The network was trained following the "standard conjugate gradient descent back-propagation" algorithm. BPN learning was realized over 2,000 epochs. The learning coefficient was set to 0.2, and the momentum equaled 0.9. Each input and output value was scaled between 0 and 1 by using the equation:

$$x' = \frac{x - x_{min}}{x_{max} - x_{min}}$$

Here x' is the normalized value, x is any one of the descriptor vectors, and x_{max} and x_{min} are the maximum and minimum values of the descriptor vector in the dataset.

In the process, the architecture of BPN was optimized; the number of hidden layer neurons varied from 5 to 18. Finally, the optimized neural network architecture was 28–13–1. In order to avoid overfitting, the best number of training epochs was obtained to achieve the smallest value of root-mean-squared error (RMSE) on an independent validation set. Training ceased at this point and the corresponding network was taken as reference. A restart strategy was used to train the network in order to avoid the local optimal solution. For the total training set (the training set used to construct BPN and an independent validation set), $R^2 = 0.876$, RMSEF = 0.536.

6.4.7 SVR model

A SVR analysis was established using the same 28 descriptors as input variables of 180 druglike molecules. Similar to other multivariate statistical models, the performance of the SVR model is related to dependent and independent variables as well as the combination of parameters used in the model. In SVR modeling, some parameters (e.g., the type and parameters of the kernel function, the regularization parameter C, and ε-insensitive loss function) have to be selected. The optimal value of the regularization parameter C is an important parameter because of its possible effects on both trained and predicted results, inasmuch as it controls the trade-off between maximizing the margin and minimizing the training error. Usually, C is an unknown parameter before modeling. If C is too small, insufficient stress will be placed on fitting the training data. If C is too large, the algorithm will overfit the training data. Therefore, C should be optimized together with the kernel functions. The optimal value of ε is also an important parameter, which depends on the type of noise present in the data and it is usually unknown.

There is a practical consideration of the number of resulting support vectors, even if enough knowledge of the noise is given to select an optimal ε. It prevents the entire training set from meeting boundary conditions, so we have to optimize ε and seek the optimal value. To obtain the best combination of parameters used in the model, a particle swarm optimization algorithm was used to achieve the smallest value of root-mean-squared error on the validation set. Particle swarms explore the search space using a population of individuals, each with a single, initially random location and velocity vector. The particle then "flies" over the state space, remembering the best solution encountered. Fitness is determined by an application-specific objective function. In each iteration, the velocity of each particle is adjusted on the basis of its momentum and the influence of the best solutions encountered by itself and its neighbors. The particle then moves to a new position and the process is repeated for a prescribed number of iterations. The particle swarm optimization algorithm as a global optimization algorithm can effectively seek the optimal value of all parameters without falling into the local optimal solution. In this work, the Gaussian kernel function was used and the optimal value of δ set to 0.0026; the values of regularization factor C and ε were set to 28.66 and 0.082, respectively. For the training set, $R^2 = 0.8837$, RMSEF = 0.520, and for the cross validation set, $Q^2 = 0.781$, RMSECV = 0.715.

Table 6.2 summarizes the performance of three models used to predict the aqueous solubility of druglike molecules. As can be shown, all three models obtain satisfactory prediction results, indicating that 28 selected molecular descriptors can effectively model the aqueous solubility of druglike molecules. The predictive results of three methods are listed in Table 6.3. One can infer that the results of SVR seem better than those of PLS and BPN with the same 28 molecular descriptors. So, SVR is more suitable to predict the aqueous solubility of druglike molecules. We mainly focus on the predictive results of SVR.

Table 6.2 The Predictive Results of the Aqueous Solubility of Druglike Molecules Using Three Different Methods

	Training Set		CV Set		Test Set	
	RMSEF	R^2	RMSECV	Q^2	RMSET	RT^2
PLS	0.599	0.846	0.737	0.767	0.769	0.708
BP	0.536	0.876	—	—	0.789	0.692
SVR	0.520	0.884	0.715	0.781	0.731	0.735

Table 6.3 The Experimental and Predicted Aqueous Solubility Values for Training and Test Set Using Three Methods

No.	Compounds	Exp	PLS	BPN	SVR
1	2,4,5-Trichlorophenol	3.079	2.199	2.514	3.436
2	2,4-DB	1.663	1.805	2.065	1.752
3	2,6-Dibromoquinone-4-chlorimide	1.77	2.286	2.178	2.129
4	2-Cyclohexyl-4,6-dinitrophenol	1.168	1.332	0.5682	1.280
5	2-Ethyl-1-hexanol	2.944	3.718	3.546	3.839
6	3,4-Dinitrobenzoic acid	3.826	3.632	2.770	3.699
7	4-Amino-2-sulfobenzoic acid	3.477	3.503	3.533	3.602
8	Acetamide	6.352	5.540	5.691	5.752
9	Acetamiprid	3.623	3.339	3.725	3.253
10	Acetanilide	3.806	3.600	3.752	3.680
11	Acetazolamide	2.991	3.045	2.620	3.170
12	Acetochlor	2.348	2.362	2.382	2.467
13	Acetylacetone	5.221	4.981	5.139	5.285
14	Acibenzolar-S-methyl	0.887	2.079	1.586	1.860
15	Acrylamide	5.806	5.011	4.849	4.738
16	Acylonitrile	4.872	5.003	5.133	4.746
17	Adenine	3.013	4.027	3.595	4.088
18	Adipic acid	4.414	4.812	4.805	4.621
19	Alanine	5.214	5.532	5.065	5.341
20	Aldicarb	3.78	3.244	3.822	3.6537
21	Allobarbital	3.258	3.706	3.392	3.384
22	Alochlor	2.38	2.246	1.763	2.153
23	Alpha-acetylbutyrolactone	5.301	4.520	4.798	4.421
24	Aminopromazine	-0.239	0.7240	0.6424	0.9926
25	Amitraz	0	−0.1744	−1.297	−0.125
26	Amobarbital	2.78	2.789	2.880	2.905
27	Ancymidol	2.813	3.077	2.962	2.8935
28	Aniline	4.556	4.602	4.990	4.703
29	ANTU	2.778	1.788	1.632	1.967
30	Arabinose	5.698	5.455	5.123	5.5717
31	Ascorbic acid	5.522	4.840	5.059	4.9405
32	Aspartic acid	3.912	4.707	4.106	4.598
33	Asulam	3.699	2.662	3.220	3.322
34	Azidamfenicol	4.301	4.242	4.086	4.174

(*Continued*)

Table 6.3 The Experimental and Predicted Aqueous Solubility Values for
Training and Test Set Using Three Methods (Continued)

No.	Compounds	Exp	PLS	BPN	SVR
35	Azintamide	3.699	3.621	2.987	3.584
36	Badische acid	2.775	3.087	2.549	2.948
37	Barban	1.042	1.497	0.9624	1.2481
38	Barbital	3.873	3.805	3.637	3.747
39	Bendiocarb	2.415	2.853	2.459	2.724
40	Benzidine	2.505	2.488	2.304	2.379
41	Bifenox	−0.397	-0.300	−0.01306	−0.07917
42	Bifentrhin	−1	−1.267	−1.171	−1.125
43	Capric acid	1.791	0.9977	2.117	1.665
44	Caproic acid	4.012	4.552	3.828	4.597
45	Carbofuran	2.505	2.664	3.213	2.379
46	Carbosulfan	−0.522	−0.3201	−0.8818	−0.3953
47	Carfentrazone-ethyl	1.343	1.546	1.687	1.330
48	Carisoprodol	2.477	2.414	3.484	3.923
49	Carmustine	3.602	3.298	3.303	3.269
50	Carnosine	4.914	3.715	3.484	3.919
51	1,6-Cleve's acid	3	2.575	2.204	2.577
52	Crotonic acid	4.934	4.775	4.685	4.473
53	Cumic acid	2.179	2.395	2.699	2.337
54	Cyanazine	2.233	2.731	2.838	2.358
55	Cyanuric acid	3.301	4.421	3.247	3.460
56	Cyclizine	3	3.768	3.462	3.125
57	Cyclobarbital	3.204	2.390	2.662	2.297
58	Cycloleucine	4.698	4.717	4.871	3.925
59	Cyproconazole	2.146	1.758	1.788	2.520
60	Cyprodinil	1.114	1.547	1.208	1.239
61	Cystine	2.049	2.360	2.307	2.175
62	Dehydroacetic acid	2.839	3.424	4.109	3.769
63	Dexamethasone	1.949	1.998	2.145	2.074
64	Dicamba	2.92	2.717	2.582	2.643
65	Dichlobenil	1.327	2.675	2.878	2.551
66	Dichlofenthion	−0.61	0.01417	−0.5020	0.001842
67	Diclofop-methyl	−0.096	0.1820	−0.2542	0.03034
68	Difenoconazole	1.177	0.5009	0.6657	0.3309
69	Digallic acid	2.699	2.397	2.818	2.522
70	Dimethenamid	3.079	2.906	2.816	2.326
71	Dimethirimol	3.079	2.863	2.389	2.591
72	EPTC	2.574	2.905	3.437	3.171

Table 6.3 The Experimental and Predicted Aqueous Solubility Values for Training and Test Set Using Three Methods (Continued)

No.	Compounds	Exp	PLS	BPN	SVR
73	Equilin	0.15	0.9589	1.020	1.044
74	Ethinamate	3.398	2.581	3.133	3.066
75	Ethirimol	2.301	2.965	2.771	2.700
76	Ethofumesate	1.699	2.197	1.494	2.056
77	Ethohexadiol	4.623	4.582	4.128	4.666
78	Ethoprop	2.875	2.888	2.684	2.749
79	Etofenprox	−3	−2.225	−2.302	−1.667
80	Fenbufen	0.344	1.375	1.082	1.062
81	Fenoxaprop-ethyl	−0.046	−0.3200	−0.1613	−0.1661
82	Fenpiclonil	0.682	1.295	0.7995	1.278
83	Fludrocortisone	2.146	2.676	2.188	1.914
84	Flufenacet	1.748	1.622	1.404	1.535
85	Flumioxazin	0.253	0.8588	1.827	0.6722
86	Fluspirilene	1	1.528	1.192	1.125
87	Fluthiacet-methyl	−0.07	0.3149	0.07576	−0.196
88	Fumaric acid	3.845	4.180	4.233	3.856
89	Furazolidone	1.603	1.712	1.829	1.952
90	Ganciclovir	3.633	3.070	3.348	3.291
91	Gluconolactone	5.77	4.935	5.019	5.044
92	Glutamic acid	3.933	4.360	4.078	4.401
93	Glycine	5.396	5.353	5.478	5.522
94	Glyphosate	4.079	4.303	3.723	4.288
95	Guaifenesin	4.698	4.007	4.627	4.359
96	Haloperidol	1.147	0.3808	0.4985	0.7246
97	Heptabarbital	2.398	2.102	2.264	2.271
98	Hexazinone	4.519	3.526	4.152	3.323
99	Histidine	4.658	4.192	3.927	4.257
100	Hydrocortisone	2.505	2.108	2.405	2.6313
101	Hydroflumethiazide	2.477	2.194	1.641	2.368
102	Hydroquinone	4.857	4.513	4.849	4.514
103	Hydroxyphenamate	4.397	3.487	3.938	3.457
104	Hydroxyproline	5.557	4.996	4.813	4.807
105	Hymexazol	4.929	4.882	6.2147	5.100
106	Idoxuridine	3.301	3.710	3.218	3.778
107	Imazapyr	4.053	3.104	3.482	3.288
108	Imazaquin	1.955	1.604	1.806	1.745
109	Imazethapyr	3.146	3.167	2.966	3.020

(Continued)

Table 6.3 The Experimental and Predicted Aqueous Solubility Values for
Training and Test Set Using Three Methods (Continued)

No.	Compounds	Exp	PLS	BPN	SVR
110	Iridomyrmecin	3.301	3.806	3.982	3.679
111	Isoflurophate	4.187	5.362	4.735	5.291
112	Isoleucine	4.536	3.952	3.622	3.788
113	Isoniazid	5.146	5.009	4.997	4.560
114	Isophorone	4.079	4.179	3.957	4.291
115	Ketanserin	1	1.158	0.6307	1.126
116	Khellin	3.017	3.328	3.449	3.205
117	Linuron	1.876	2.223	2.080	2.329
118	Methomyl	4.763	4.159	3.748	4.033
119	p-Fluorobenzoic acid	3.079	3.579	3.839	3.006
120	Phthalazine	4.698	4.896	4.440	4.734
121	Phthalic acid	3.846	3.468	3.257	3.293
122	Phthalimide	2.556	3.658	3.521	3.417
123	p-Hydroxybenzoic acid	3.699	3.749	3.765	2.811
124	Picloram	2.633	2.784	2.980	2.796
125	Picric acid	4.103	3.514	2.796	3.430
126	Thionazin	3.057	2.961	2.759	3.022
127	Benzocaine	2.898	3.044	3.352	3.254
128	Theophylline	3.886	4.068	3.408	3.805
129	Antipyrine	5.665	4.059	4.521	3.690
130	Nitrofurantoin	1.997	2.209	2.107	2.162
131	Phenytoin	1.412	2.773	2.530	2.382
132	Testosterone	1.39	1.272	1.941	1.516
133	Lindane	0.864	1.662	0.8797	0.9902
134	Phenolphthalein	2.603	1.644	1.939	1.348
135	Malathion	2.159	1.667	1.407	2.032
136	Diazepam	1.699	1.106	0.800	1.082
137	Aspirin	3.663	2.809	2.883	2.843
138	1-Naphthol	3.1761	3.196	3.382	3.1273
139	2-Amino-5-bromobenzoic-acid	2.2601	2.867	2.757	2.800
140	4-Iodophenol	3.6284	3.502	3.305	3.502
141	5-Bromo-2-4-dihydroxybenzoic acid	2.7474	3.014	2.644	2.621
142	Acetaminophen	4.1139	4.145	4.288	4.240
143	Acetazolamide	2.9117	2.773	2.553	2.955
144	Alprenolol	2.7634	2.777	2.753	2.886
145	Amantadine	3.3263	3.509	3.402	3.322

Table 6.3 The Experimental and Predicted Aqueous Solubility Values for Training and Test Set Using Three Methods (Continued)

No.	Compounds	Exp	PLS	BPN	SVR
146	Amitryptyline	0.89209	0.8680	0.8439	0.7658
147	Amodiaquine	−0.24413	−0.1396	−0.026	−0.1186
148	Atropine	3.4594	2.987	3.053	3.045
149	Azathioprine	2.2355	0.8893	1.832	2.147
150	Benzylimidazole	2.9415	2.969	3.272	3.06
151	Bromogramine	1.3483	2.168	1.529	1.831
152	Bupivacaine	2.2355	1.785	1.729	1.265
153	Carprofen	0.74036	0.02640	0.4602	0.05476
154	Carvedilol	1.3541	0.6887	1.044	1.228
155	Cephalothin	2.6599	2.423	2.309	2.533
156	Chlorpheniramine	2.7709	2.282	2.186	2.025
157	Chlorpromazine	0.43136	0.06481	−0.2206	0.1872
158	Chlorprothixene	−0.36653	−0.8919	−0.7599	−0.4836
159	Cimetidine	3.7101	3.674	3.873	3.724
160	Ciprofloxacin	1.9243	2.861	2.677	2.444
161	Danofloxacin	2.6536	2.272	2.492	2.242
162	Deprenyl	2.7597	2.573	2.792	2.485
163	Desipramine	1.7993	1.013	0.9585	1.144
164	Diazoxide	2	2.008	1.944	1.948
165	Diclofenac	0.01536	0.09516	−0.2512	0.1417
166	Difloxacin	2	1.762	1.537	1.873
167	Diltiazem	2.4579	2.256	3.079	2.398
168	Diphenydramine	2.4609	3.487	3.133	2.630
169	Diphenylhydantoin	1.5441	2.422	2.330	2.199
170	Enrofloxacin	2.3747	2.765	2.899	2.286
171	Famotidine	2.8808	1.998	2.125	2.632
172	Fenoprofen	1.6812	1.624	1.362	2.222
173	Flufenamic acid	0.093422	0.5372	0.1199	0.4487
174	Flumequi	1.6812	2.276	2.338	1.463
175	Flurbiprofen	1.2355	1.726	1.397	1.393
176	Glipizide	0.16137	0.6512	0.3535	0.3867
177	Guanine	0.74819	3.640	3.266	3.574
178	Hexobarbital	2.699	2.733	2.570	2.734
179	Hydroflumethiazide	2.5563	2.438	1.838	2.430
180	5-Hydroxybenzoic acid	3.6758	3.692	3.819	3.635
181	Ibuprofen	1.716	1.913	2.052	1.930
182	Lomefloxacin	3.2125	3.025	2.851	2.846

(Continued)

Table 6.3 The Experimental and Predicted Aqueous Solubility Values for
Training and Test Set Using Three Methods (Continued)

No.	Compounds	Exp	PLS	BPN	SVR
183	Maprotiline	0.74819	3.321	1.524	0.7131
184	Mefenamic acid	−1.3546	0.6942	0.4065	0.6415
185	Metoclopramide	1.9138	3.542	3.040	3.284
186	Metronidazole	4.012	4.465	4.124	3.920
187	Miconazole	0.54407	0.3580	0.3381	0.4357
188	Nalidixic acid	1.7559	2.064	2.392	2.180
189	Naloxone	2.617	2.946	2.616	2.742
190	Naproxen	0.86332	1.704	1.376	0.522
191	Niflumic acid	0.8451	1.361	0.8581	0.9707
192	Nitrofurantoin	2.1367	2.465	2.748	2.665
193	Norfloxacin	2.7474	2.508	2.660	2.848
194	Nortriptyline	1.3979	0.8131	1.012	0.8675
195	Ofloxacin	4.2923	3.155	2.593	2.937
196	Oxytetracycline	2.5798	2.487	2.355	2.705
197	Papaverine	1.6628	1.582	2.062	1.707
198	Phenantroline	3.6377	2.992	2.786	2.882
199	Phenazopyridine	1.1367	1.480	1.543	1.563
200	Phenobarbital	3.0719	2.402	2.498	2.178
201	Phenylbutazone	1.0983	1.238	1.141	1.224
202	Phthalic acid	3.616	3.755	3.804	3.489
203	Pindolol	1.6021	2.683	2.702	2.688
204	Piroxicam	0.716	2.112	1.802	1.828
205	Procaine	3.6532	3.327	3.839	3.371
206	Propranolol	1.9191	2.580	2.191	2.409
207	Quinine	2.7243	2.073	2.218	2.114
208	Ranitidine	2.9956	3.330	2.912	3.030
209	Sarafloxacin	2.4472	1.812	1.706	1.939
210	Sparfloxacin	2.2227	2.941	3.008	2.710
211	Sulfacetamide	3.8109	3.493	3.741	3.684
212	Sulfamethazine	2.7059	2.808	2.528	2.581
213	Sulfasalazine	−0.5376	−0.4970	0.2836	−0.4116
214	Sulfathiazole	2.7177	2.456	2.365	2.608
215	Sulindac	1.0414	1.546	1.339	1.166
216	Tetracaine	2.4116	2.726	2.257	2.538
217	Tetracycline	2.7218	2.845	2.518	2.595
218	Thymol	2.9908	3.215	2.981	3.102
219	Tolmetin	1.3222	1.702	1.514	1.492
220	Trichloromethiazide	2.0531	1.243	1.054	1.741

Table 6.3 The Experimental and Predicted Aqueous Solubility Values for Training and Test Set Using Three Methods (Continued)

No.	Compounds	Exp	PLS	BPN	SVR
221	Trimethoprim	2.5119	3.468	2.7349	3.2238
222	Trimipramine	0.68124	0.5626	0.51238	0.80744
223	Tryptamine	1.9031	2.837	3.2316	3.0801
224	Verapamil	1.6821	1.079	0.55845	1.5558
225	Warfarin	0.70757	1.081	1.28	1.1038

6.4.8 Applicability domain and external validation

It is well known that QSPR is based on an assumption that compounds of similar structure will exhibit similar properties. Moreover, the model built by some training set will be strongly dependent upon the structures defined by the training set. If there are some special chemicals called outliers departing from the bulk of the dataset, they will destroy the similarity of the chemicals and influence the fitting and subsequent prediction ability of the QSPR model. So the detection of outliers in the training set is very essential to generate a robust and reliable model. This is also consistent with the applicability domain criterion in the OECD principles.

To obtain a robust model, a Monte Carlo cross-validation method (MC), developed by our group (Cao et al. [59]), was used to find the outliers in the training samples. In a QSPR study, if we model given QSPR data by a single training/test set division, then we can obtain the predictive errors of this test set, characterizing the behavior of these samples within the current division. However, because these results of the current division highly depend on the way that we sample the data, different training/test data division should yield different results about predictive errors. Thus, by different training/test data division by means of Monte Carlo, we can obtain a large number of QSPR models and corresponding predictive errors to get some insight into the data structure statistically. The MC method is based on such an assumption and aims to study the distribution of predictive errors and then reflect sample behavior. The MC method can inherently provide a feasible way to detect different kinds of outliers by establishment of many cross-predictive models. With the help of the distribution of predictive errors thus obtained, it seems to be able to reduce the risk caused by the masking effect.

In addition, a new display is proposed in which the absolute values of the mean value of predictive errors are plotted versus standard deviations of predictive errors. The plot divides the data into normal

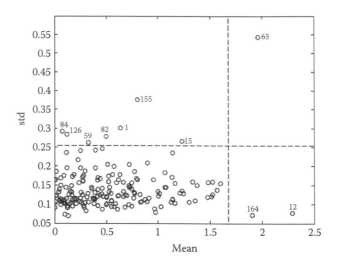

Figure 6.3 Standard deviation of predictive errors versus mean of predictive errors on the training set using the MC method. Two broken lines split the plot into four parts: the top left corner (**X** outliers), the lower right corner (**y** outliers), the top right corner (abnormal samples), and the lower left corner (normal samples).

samples, *y*-direction outliers and **X**-direction outliers. Thus this method can simultaneously detect **X** outliers and y outliers. Figure 6.3 shows the results of outlier detection using MC. It can be seen from Figure 6.3 that two broken lines split the plot into four parts: the top-left corner (**X** outliers), the lower-right corner (**y** outliers), and the top-right corner (abnormal samples), and the lower-left corner (normal samples). Thus, samples 65, 12, and 164 are considered as **y** outliers and samples 65, 155, 1, 84, 126, 82, 59 and 15 are considered as **X** outliers. In addition, these **X** outliers except for 65 have smaller predictive errors, which indicate that these samples can be seen as "good leverage" chemicals. Sample 65 is a special point that is both an **X** and a **y** outlier. So, this sample can be seen as a "bad leverage" point. When **y** outliers are deleted, the remaining 177 druglike molecules are used to re-establish the SVR model. The model gives the R^2 value of 0.898, RMSEF value of 0.476 for the training set, and Q^2 value of 0.803, and RMSECV value of 0.66 for the cross-validation set, respectively. (For PLS: RMSEF = 0.559, R^2 = 0.858, RMSECV = 0.698, Q^2 = 0.783; for BPN: RMSEF = 0.525, R^2 = 0.879.) Compared with the aforementioned SVR model, the RMSEF and RMSECV values for both the training set and cross-validation set reduced largely, and at the same time the R^2 and Q^2 values increased by 0.015 and 0.02, respectively. Figure 6.4 shows the predictive results of SVR.

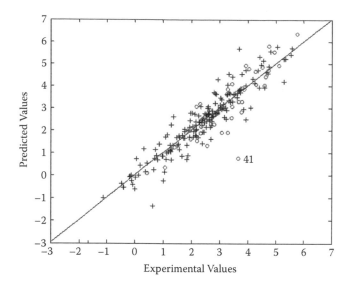

Figure 6.4 Predictive log S values versus experimental values for the training set (plus) and the test set (circle) by SVR.

An independent test set of 45 druglike molecules was used to test the predictive model. The predictive results for the test set are also shown in Table 6.3 and Figure 6.4. The correlation coefficients (RT^2) and root mean squared error of testing (RMSET) of three models are 0.708 and 0.769 for PLS, 0.692 and 0.789 for BPN, and 0.735 and 0.731 for SVR, respectively. It is easy to find that SVR could obtain better predictive results for the test set. To check whether the test samples lie in the scope of the training sample space, the commonly used Mahalanobis distance (MD) is adopted to detect the outliers in the **X**-direction. Unlike the Euclidean distance, MD takes the effect of the data distribution into account and can better describe the underlying structure of the data than Euclidean distance. So, we computed the MD of test samples based on 177 training samples. No test samples were found to have a distance greater than 2 times (4.436) the mean distance of the test samples from the training set. So, we could conclude that the test set did not include X outliers. However, as can be seen in Figure 6.4, sample 41 has a significantly large residual and could be seen as a y outlier. A possible reason might be that the structure of sample 41 is not similar enough to those of the training set or the selected descriptors are not adequate to describe the sample. With this sample removed, RT^2 and RMSET based on SVR remarkably increased (RT^2 = 0.816 and RMSET = 0.596). From the above results, the models we constructed not only have good fitting ability but also high capability for assessing the external validation set.

6.4.9 Model interpretation

It is shown in Table 6.3 that the prediction results of the three methods are good and acceptable, which indicates 28 molecular descriptors selected can effectively represent drug aqueous solubility. To compare and interpret every molecular descriptor well, the variable importance of molecular descriptors is computed based on SVR. At each round, one molecular descriptor is removed from the total pool of molecular descriptors. The remaining 27 molecular descriptors are used to build the SVR model and a new RMSECV is computed. The difference between the new RMSECV and the RMSECV in Table 6.2 can be seen as the measure of variable importance of the removed molecular descriptor. Thus, we repeat the process 28 times to obtain the variable importance of every molecular descriptor.

Figure 6.5 shows the variable importance of 28 molecular descriptors based on SVR. From Figures 6.2 and 6.5 and Table 6.1, we can clearly see that variables ALOGP2 (squared Ghose–Crippen octanol–water partition coefficient (log P^2)) and ALOGP (Ghose–Crippen octanol–water partition coefficient (log P)) are two important factors that influence drug aqueous solubility. Log P is frequently used to estimate membrane permeability and the bioavailability of compounds for an orally administered drug must be lipophilic enough to cross the lipid bilayer of the membranes and sufficiently water soluble to be transported in the blood and the lymph. Log P is frequently used in quantitative structure–property relationships

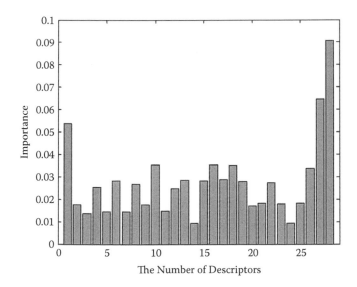

Figure 6.5 The variable importance of 28 molecular descriptors based on SVR.

as a measure of the lipophilic character of molecules. In PLS regression, we can see that the coefficients of ALOGP2 and ALOGP are negative. Negative values in the regression coefficients indicate that the greater the value of the descriptor, the lower the value of log S, whereas positive values indicate that the greater the value of the descriptor, the higher the value of log S. Among the descriptors, some of them (e.g., ATS1m, MATS2m, MATS7m, MATS3e, BELe4, VRp2, RGyr, RDF020m, RDF065m, RDF015e, RDF100e, Mor27p, HATS8v, R3v, R4e) are weighted by molecular mass, atomic polarizabilities, atomic Sanderson electronegativities, and atomic van der Waals volumes. These factors might partly influence drug aqueous solubility.

Molecular volume determines the transport characteristics of molecules, such as intestinal absorption or blood–brain barrier penetration. Volume is therefore often used in QSPR studies to model molecular properties and biological activity. The steric effects, characterizing bulk properties of a molecule, can be described with molecular volume. Molecular volume is clearly an important descriptor for aqueous solubility. In order for a solute to enter into an aqueous solution, a cavity must be formed in the solvent for the solute molecule to occupy. Water as a solvent would much prefer to interact with itself or other hydrogen bonding or ionic species than with a nonpolar solute, so there is an increasing penalty for larger solutes. Increasing molecular volume leads to increasing cavity formation energy in water: the larger the solute is, the greater the energy demand to make the cavity and the lower the solubility. Molecular polarizabilities and Sanderson electronegativities are also important parameters. As we know, the more similar the polarities of the compounds are, the more easily the compounds mutually dissolve. As molecular polarizabilities and Sanderson electronegativities increase, drug aqueous solubility also increases. Moreover, some common function groups of the drug molecules play an important role in the prediction of the compounds' properties, such as C-026, C-040, H-052, and O-058.

6.5   Support vector machines for classification

As we discussed in Chapter 3, the kernel function used in SVM can be further extended to a large class of methods. Here, a two-step nonlinear classification algorithm is proposed to model the structure–activity relationship (SAR) between bioactivities and molecular descriptors of compounds, which consists of kernel principal component analysis (KPCA) and linear support vector machines (LSVM). KPCA is used to remove some uninformative gradients such as noise and then exactly capture the latent structure of the training dataset using some new variables called the principal components in the kernel feature space. The linear support vector machine (LSVM) makes full use of the maximal margin hyperplane to

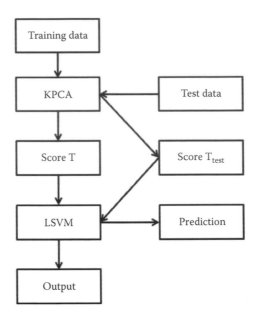

Figure 6.6 The flow chart of the KPCA-LSVM algorithm. T indicates the scores matrix of KPCA. In the KPCA-LSVM algorithm, there are some important parameters that need to be further optimized (see Section 6.5.4).

give the best generalization performance in the KPCA-transformed space. The combination of KPCA and LSVM can effectively improve the prediction performance compared to the original SVMs.

6.5.1 *Two-step algorithm: KPCA plus LSVM*

For nonlinear classification, the original SVMs first project the input feature vectors into a high-dimensional feature space using a kernel function $K(\mathbf{x}_i, \mathbf{x}_j)$ and then perform LSVM in the kernel feature space. However, the variables in the kernel feature space may include some redundant information or noise that may affect the prediction accuracy of the established model. It is imperative to remove such useless information before performing LSVM. In order to deal with such a situation, a two-step nonlinear algorithm based on the combination of KPCA and LSVM (KPCA-LSVM) was developed. The two-step KPCA-LSVM algorithm is given below (see Figure 6.6).

> **Step 1: Perform KPCA in input space.** KPCA is carried out in the input space to extract the compact underlying structure of the dataset. We can calculate its orthonormal eigenvectors $\mathbf{V}_k = [\mathbf{v}_1, \mathbf{v}_2, ..., \mathbf{v}_k]$ corresponding to k largest nonzero eigenvalues $\lambda_1, \lambda_2, ..., \lambda_k$. So, the

corresponding scores can be computed $T_k = \varphi(X) \cdot V_k$ and the original kernel matrix can be reconstructed as $K = [\varphi(X)] \, [\varphi(X)]^t \approx [T_k] \, [T_k]^t$. Here, k should be further optimized by means of model selection techniques such as cross-validation.

Step 2: Perform LSVM in KPCA-transformed space. The LSVM can be directly carried out by means of the reconstructed kernel matrix in the KPCA-transformed space.

Here, some remarks on the above algorithm should be given. The KPCA-LSVM method has a consistent framework with the existing non-linear SVMs. KPCA-LSVM will be changed into the original nonlinear SVMs when all the scores in KPCA are used to perform the LSVM algorithm. However, the KPCA-LSVM algorithm is more flexible compared to the existing nonlinear SVMs, especially when the variables in the kernel-defined feature space include some redundant information or noise. In addition, the primal PCA algorithm may destroy the underlying nonlinear structure possessed by the training dataset. However, KPCA is performed in the kernel-defined feature space and is more suitable to capture the underlying nonlinear structure of the training dataset. So, KPCA-LSVM is theoretically more reasonable compared with commonly used SVM coupled with PCA.

6.5.2   Dataset description

In order to illustrate the performance of the proposed approach, three examples often used for comparison in the SAR literature were used to evaluate this KPCA-LSVM approach. The three datasets, Pgp, TdP, and HIA, were obtained from Xue et al. [41]. Pgp is a transmembrane protein capable of transporting a wide variety of anticancer drugs out of the cell, resulting in an obstacle in chemotherapeutic treatment. An increased expression of Pgp is associated with multidrug resistance (MDR). Many studies have been undertaken to develop MDR-reversing compounds with potential clinical significance. A total of 116 substrates and 85 non-substrates of Pgp were collected in the dataset of Pgp. TdP is a potentially fatal polymorphic ventricular tachycardia. It may also be induced as an adverse effect of drugs that cause QT prolongation. This effect is present in different categories of therapeutic agents, for example, antihistamines, antidepressants, or macrolide antibiotics. The TdP dataset included 85 TdP-inducing agents (TdP+) and 276 noninducing compounds (TdP–). Moreover, the absorption of a drug compound through the human intestinal cell lining is an important property for potential drug candidates. There are 196 HIA compounds, including 131 absorbable (HIA+) and 65 nonabsorbable (HIA–) compounds, classified by the "measured absorption rate" of the 70% criterion. In this work, we used the original set of

159 descriptors provided by Xue et al. for the HIA, Pgp, and TdP sets to represent these molecules.

6.5.3   Performance evaluation

Because of the small size of the SAR datasets, we therefore used fivefold cross-validation to estimate the accuracy of our models. For fivefold cross-validation, the training set was split into five roughly equal-sized parts first, and then the model was fit to four parts of the data and the error rates of the other part calculated. The process was repeated five times so that every part could be predicted as a validation set. The parameters employed to evaluate the behavior in this investigation were some commonly used ones in classification problems: true positives (TP), true negatives (TN), false positives (FP), and false negatives (FN). There were several criteria for assessing prediction performance including sensitivity SE (the prediction accuracy of active compounds), specificity SP (the prediction accuracy of inactive compounds), overprediction accuracy (R), and Matthews correlation coefficient (MCC), which are given by the equations:

$$SE = \frac{TP}{TP + FN}$$

$$SP = \frac{TN}{TN + FP}$$

$$R = \frac{TP + TN}{TP + FP + TN + FN}$$

$$MCC = \frac{TP \times TN - FN \times FP}{\sqrt{(TP + FN)(TP + FP)(TN + FN)(TN + FP)}}$$

where TP is the number of positive samples predicted correctly, TN is the number of negative samples predicted correctly, FP is the number of negative samples predicted as positive, and FN is the positive samples predicted as negative.

6.5.4   Effects of model parameters

In our study, the Gaussian kernel function is used to construct the nonlinear mapping. The Gaussian kernel function can be represented by

$$K(\mathbf{x}_i, \mathbf{x}_j) = \exp(-||\mathbf{x}_j - \mathbf{x}_i||^2 / 2\delta^2)$$

which has been extensively used in different studies with good performance. Thus, in KPCA-LSVM, there are three very important parameters, including k (the number of principal components), C (the regularization parameter), and δ (the width of the Gaussian kernel function), which need to be further optimized. The number k of the principal components controls the ability of reconstructing the dataset in the kernel-defined feature space. The choice of a suitable k value depends on the contribution to the response values and should be further optimized by means of model selection techniques such as cross-validation and the like. The optimal value of the regularization parameter C is an important parameter because of its possible effects on both trained and predicted results. Specific discussions can be seen in Section 6.4.7. The width δ of the Gaussian kernel function is also an important parameter that needs to be tuned. A very small δ can excessively model the local structure of the training data and so may overfit the training data, and a high δ does not capture the underlying structure of the training data and so may underfit the training data. Generally speaking, these parameters of KPCA-LSVM are mutually interrelated and thus are optimized jointly. In the study, a multiparameter grid search strategy can be applied to seek the optimal combination of model parameters simultaneously.

6.5.5   Prediction results for three SAR datasets

The prediction accuracies of KPCA-LSVM for three SAR datasets were primarily evaluated by means of fivefold cross-validation. All model parameters were further optimized via a grid search strategy. For the regularization parameter C, we set five values ($C = 1, 2, 5, 10, 100$). For the number k of the principal components, it ranged from 1 to 60. For the width δ of Gaussian kernel function, we first estimated a suitable range and then set seven values ($\delta = 0.0001, 0.0002, 0.0003, 0.0005, 0.001, 0.005, 0.01$). Thus, we made use of $5 \times 60 \times 7 = 2{,}100$ grid points to search for the optimal combination of model parameters. For two methods, each molecular descriptor for three SAR datasets was normalized to have zero mean and unit variance.

Table 6.4 shows the comparison results of the prediction accuracies on three SAR datasets from two different nonlinear classification methods. As shown in Table 6.4, SVM uniformly achieves the worst prediction performance on three real SAR datasets, together with mean value 77.1% of prediction accuracy, which indicates that SVM have poor prediction ability. A possible reason for this is that the redundant information or noise in the molecular descriptors seriously influences the prediction performance of the SVM. Compared with the SVM, KPCA-LSVM significantly improves the prediction ability on the three SAR datasets, together with mean value 81.4% of prediction accuracy. For datasets HIA and Pgp, we

Table 6.4 The Results of Two Methods Used on Three Datasets

	SVM				KPCA-LSVM			
Datasets	SP(%)	SE(%)	R(%)	MCC(%)	SP(%)	SE(%)	R(%)	MCC(%)
HIA	60.0	83.3	77.6	48.0	67.7	90.8	83.2	61.0
P-gp	64.7	77.6	72.1	42.6	65.9	87.1	78.1	54.7
TdP	82.6	78.8	81.7	55.9	81.3	88.3	82.8	61.7
Average	**69.1**	**79.9**	**77.1**	**48.8**	**71.6**	**88.7**	**81.4**	**59.1**

Note: The optimal parameters used for KPCA-LSVM: for HIA dataset, $\delta = 0.0001$, $C = 5$, $k = 52$. For P-gp dataset, $\delta = 0.005$, $C = 1$, $k = 44$. For TdP dataset, $\delta = 0.0002$, $C=5$, $k = 47$.

can see from Table 6.4 that the prediction accuracies of KPCA-LSVM are far superior to ones from SVM. For the HIA dataset, the prediction accuracy increases from 77.6% of the SVM to 83.2% of KPCA-LSVM, which is a 5.6% improvement in prediction ability. For the Ppg dataset, the prediction accuracy increases from 72.1% of SVM to 78.1% of KPCA-LSVM, which is a 6% improvement in the prediction ability. However, for the TdP dataset, the improvement in the prediction accuracy is not significant (for SVM: 81.7%, for KPCA-LSVM: 82.8%). For all three SAR datasets, the prediction ability of KPCA-LSVM is completely superior to ones from the SVM, which seems to indicate that KPCA-LSVM indeed outperforms SVM on the three SAR datasets by means of the use of KPCA. KPCA-LSVM makes full use of KPCA to carry out the dimensionality reduction or denoising in the kernel-defined feature space and thus remarkably improve the prediction performance of subsequent LSVM. It can be concluded that the classification models we construct using the KPCA-LSVM method should be better than the original SVM method.

References

1. Crum-Brown, A. 1864. On the theory of isomeric compounds. *Trans. Roy. Soc. Edinb.*, 23:7077–7719.
2. Crum-Brown, A. 1866. On an application of mathematics to chemistry. *Proc. Roy. Soc. Edinb.*, 73:89–90.
3. Korner, W. 1874. Studi sulla Isomeria delle Così` Dette Sostanze Aromatiche a Sei Atomi di Carbonio. *Gazz. Chim. Ital.*, 4:242.
4. Mills, E.J. 1884. On melting point and boiling point as related to composition. *Philos. Mag.*, 17:173–187.
5. Richet, M.C. 1893. Note sur la Rapport entre la Toxicite´ et les Proprie´ te´ s Physiques des Corps. *Compt. Rend. Soc. Biol.*, 45:775–776.
6. Wiener, H. 1947. Influence of interatomic forces on paraffin properties. *J. Chem. Phys.*, 15:766.
7. Platt, J.R. 1947. Influence of neighbor bonds on additive bond properties in paraffins. *J. Chem. Phys.*, 15:419–420.

8. Kier, L.B. 1971. New York: Academic Press.

9. Pauling, L. 1939. *The Nature of the Chemical Bond*. Ithaca, NY: Cornell University Press.

10. Sanderson, R.T. 1952. Electronegativity I. Orbital electronegativity of neutral atoms. *J. Chem. Educ.*, 29:540–546.

11. Fukui, K., Yonezawa, Y., and Shingu, H. 1954. Theory of substitution in conjugated molecules. *Bull. Chem. Soc. Jpn.* 27:423–427.

12. Mulliken, R.S. 1955. Electronic population analysis on LCAO-MO molecular wave functions. I. *J. Chem. Phys.* 23:1833–1840.

13. Hammett, L.P. 1937. The effect of structure upon the reactions of organic compounds. Benzene derivatives. *J. Am. Chem. Soc.*, 59: 96–103.

14. Hammett, L.P. 1938. Linear free energy relationships in rate and equilibrium phenomena. *Trans. Faraday Soc.* 34:156–165.

15. Taft, R.W. 1953. The general nature of the proportionality of polar effects of substituent groups in organic chemistry. *J. Am. Chem. Soc.*, 75:4231–4238.

16. Taft, R.W. 1953. Linear steric energy relationships. *J. Am. Chem. Soc.*, 75:4538–4539.

17. Hansch, C., Maloney, P.P., Fujita, T., and Muir, R.M. 1962. Correlation of biological activity of phenoxyacetic acids with Hammett substituent constants and partition coefficients. *Nature*, 194:178–180.

18. Hansch, C., Muir, R.M., Fujita, T., Maloney, P.P., Geiger, F., and Streich, M. 1963. The correlation of biological activity of plant growth regulators and chloromycetin derivatives with Hammett constants and partition coefficients. *J. Am. Chem. Soc.*, 85:2817–2824.

19. Free, S.M. and Wilson, J.W.A. 1964. Mathematical contribution to structure–activity studies. *J. Med. Chem.*, 7:395–399.

20. Balaban, A.T. and Harary, F. 1971. The characteristic polynomial does not uniquely determine the topology of a molecule. *J. Chem. Doc.*, 11:258–259.

21. Balaban, A.T.E. 1976. *Chemical Applications of Graph Theory*. New York: Academic Press.

22. Randic, M. 1974. On the recognition of identical graphs representing molecular topology. *J. Chem. Phys.*, 60:3920–3928.

23. Randic, M. 1975. On characterization of molecular branching. *J. Am. Chem. Soc.*, 97:6609–6615.

24. Kier, L.B., Hall, L.H., Murray, W.J., and Randic, M. 1975. Molecular connectivity. I: Relationship to nonspecific local anesthesia. *J. Pharm. Sci.*, 64:1971–1974.

25. Rohrbaugh, R.H. and Jurs, P.C. 1987. Descriptions of molecular shape applied in studies of structure/activity and structure/property relationships. *Anal. Chim. Acta*, 199:99–109.

26. Stanton, D.T. and Jurs, P.C. 1990. Development and use of charged partial surface area structural descriptors in computer-assisted quantitative structure–property relationship studies. *Anal. Chem.*, 62:2323–2329.

27. Todeschini, R., Lasagni, M., and Marengo, E. 1994. New molecular descriptors for 2D- and 3D-structures, theory. *J. Chemom.*, 8:263–273.

28. Ferguson, A.M., Heritage, T.W., Jonathon, P., et al. 1997. EVA: A new theoretically based molecular descriptor for use in QSAR/QSPR Analysis. *J. Comput. Aid. Mol. Des.*, 11:143–152.

29. Schuur, J., Selzer, P., and Gasteiger, J. 1996. The coding of the three-dimensional structure of molecules by molecular transforms and its application to structure-spectra correlations and studies of biological activity. *J. Chem. Inf. Comput. Sci.*, 36:334–344.

30. Consonni, V., Todeschini, R., and Pavan, M. 2002. Structure/response correlations and similarity/diversity analysis by GETAWAY descriptors. Part 1. Theory of the novel 3D molecular descriptors. *J. Chem. Inf. Comput. Sci.*, 42:682–692.

31. Goodford, P.J. 1985. A computational procedure for determining energetically favorable binding sites on biologically important macromolecules. *J. Med. Chem.*, 28:849–857.

32. Cramer, R.D.I., Patterson, D.E., and Bunce, J.D. 1988. Comparative molecular field analysis (comfa). 1. Effect of shape on binding of steroids to carrier proteins. *J. Am. Chem. Soc.*, 110:5959–5967.

33. Klebe, G., Abraham, U., and Mietzner, T. 1994. Molecular similarity indices in a comparative analysis (CoMSIA) of drug molecules to correlate and predict their biological activity. *J. Med. Chem.*, 37:4130–4146.

34. Jain, A.N., Koile, K., and Chapman, D. 1994. Compass: Predicting biological activities from molecular surface properties. Performance comparisons on a steroid benchmark. *J. Med. Chem.*, 37:2315–2327.

35. Pastor, M., Cruciani, G., McLay, I.M., Pickett, S.D., and Clementi, S. 2000. GRid-INdependent Descriptors (GRIND): A novel class of alignment-independent three-dimensional molecular descriptors. *J. Med. Chem.*, 43:3233–3243.

36. Gasteiger, J. 2003. *Handbook of Chemoinformatics.* Weinheim, Germany: Wiley-VCH.

37. Oprea, T.I. 2004. 3D QSAR *Modeling in Drug Design. ComputationalMedicinal Chemistry for Drug Discovery.* New York: Marcel Dekker, 571–616.

38. Dudek, A.Z., Arodz, T., and Galvez, J. 2006. Computational methods in developing quantitative structure-activity relationships (QSAR): A review. *High Throughput Screening,* 9:213–228.

39. Burbidge, R., Trotter, M., Buxton, B., and Holden, S. 2001. Drug design by machine learning: Support vector machines for pharmaceutical data analysis. *Comput. Chem.*, 26:5–14.

40. Xue, Y., Yap, C.W., Sun, L.Z., Cao, Z.W., Wang, J.F., and Chen, Y.Z. 2004. Prediction of P-glycoprotein substrates by a support vector machine approach. *J. Chem. Inf. Comput. Sci.*, 44:1497–1505.

41. Xue, Y., Li, Z.R., Yap, C.W., Sun, L.Z., Chen, X., and Chen, Y.Z. 2004. Effect of molecular descriptor feature selection in support vector machine classification of pharmacokinetic and toxicological properties of chemical agents. *J. Chem. Inf. Comput. Sci.*, 44:1630–1638.

42. Muller, K.-R., Ratsch, G., Sonnenburg, S., Mika, S., Grimm, M., and Heinrich, N. 2005. Classifying drug-likeness with kernel-based learning methods. *J. Chem. Inf. Model.*, 45:249–253.

43. Yao, X.J., Panaye, A., Doucet, J.P., et al. 2004. Comparative study of QSAR/QSPR correlations using support vector machines, radial basis function neural networks, and multiple linear regression. *J. Chem. Inf. Comput. Sci.*, 44:1257–1266.

44. Svetnik, V., Wang, T., Tong, C., Liaw, A., Sheridan, R.P., and Song, Q. 2005. Boosting: An ensemble learning tool for compound classification and QSAR modeling. *J. Chem. Inf. Model.*, 45:786–799.

45. Svetnik, V., Liaw, A., Tong, C., Culberson, J.C., Sheridan, R.P., and Feuston, B.P. 2003. Random forest: A classification and regression tool for compound classification and QSAR modeling. *J. Chem. Inf. Comput. Sci.*, 43:1947–1958.

46. Xue, C.X., Zhang, R.S., Liu, H.X., et al. 2004. QSAR models for the prediction of binding affinities to human serum albumin using the heuristic method and a support vector machine. *J. Chem. Inf. Comput. Sci.*, 44:1693–1700.

47. Merkwirth, C., Mauser, H., Schulz-Gasch, T., Roche, O., Stahl, M., and Lengauer, T. 2004. Ensemble methods for classification in cheminformatics.

48. Liu, Y. 2004. A comparative study on feature selection methods for drug discovery. *J. Chem. Inf. Comput. Sci.*, 44:1823–1828.

49. Byvatov, E. and Schneider, G. 2004. SVM-based feature selection for characterization of focused compound collections. *J. Chem. Inf. Comput. Sci.*, 44:993–999.

50. Helma, C., Jfa, C., et al. 2004. Data mining and machine learning techniques for the identification of mutagenicity inducing substructures and structure activity relationships of noncongeneric compounds. *Amer. Chem. Soc.*

51. Xue, C.X., Zhang, R.S., Liu, H.X., et al. 2004. An accurate QSPR study of O-H bond dissociation energy in substituted phenols based on support vector machines. *J. Chem. Inf. Comput. Sci.*, 44:669–677.

52. Xue, C.X., Zhang, R.S., Liu, M.C., Hu, Z.D., and Fan, B.T. 2004. Study of the quantitative structure-mobility relationship of carboxylic acids in capillary electrophoresis based on support vector machines. *ChemInf.*, 35.

53. Xue, C.X., Zhang, R.S., Liu, H.X., Liu MC, Hu ZD, and Fan BT. 2004. Support vector machines-based quantitative structure property relationship for the prediction of heat capacity. *J. Chem. Inf. Comput. Sci.*. 44:1267–1274.

54. Liu, H.X., Xue, C.X., Zhang, R.S., et al. 2004. Quantitative prediction of $\log_k$ of peptides in high-performance liquid chromatography based on molecular descriptors by using the heuristic method and support vector machine. *J. Chem. Inf. Comput. Sci.*, 44:1979–1986.

55. Index, T.M., Editor, 2001. *An Encyclopedia of Chemicals, Drugs, and Biologicals*. NJ: Merck & Co.

56. Kennard, R.W. and Stone, L.A. 1969. Computer aided design of experiments. *Technometrics*, 11:137–148.

57. Kutner, Nachtsheim, Neter, Editors, 2005. *Applied Linear Regression Models*. Higher Education Press.

58. Shao, J. 1993. Linear model selection by cross-validation. *J. Amer. Statist. Assoc.*, 88:486–494.

59. Cao, D.S., Liang, Y.Z., Xu, Q.S., Li, H.D., and Chen, X. 2010. A new strategy of outlier detection for QSAR/QSPR. *J. Comput. Chem.*, 31:592–602.

chapter seven

Support vector machines applied to traditional Chinese medicine

Contents

7.1   Introduction

In this chapter, we start with an introduction to traditional Chinese medicines (TCM) together with the quality control of TCM by means of the chromatography fingerprint technique. Then, we give an example to illustrate the use of support vector machines to build a class prediction model for recognition of authentic traditional Chinese medicines. Variable selection for support vector machines is emphasized and different methods are also compared.

7.2   Traditional chinese medicines
   and their quality control

Traditional Chinese medicines, also known as herbal medicine (HM) in China, include a range of traditional medicine practices mainly originating in China. To the best of our knowledge, TCMs are well accepted in the mainstream of medical care all over East Asia, such as China, Korea, Japan, and so on. TCM is also considered as an alternative medical system in the Western world. According to the literature, there are several thousands of TCMs used in traditional medicine practices. In the great monograph *Compendium of Materia Medica*, authored by Shi-Zhen Li during the

Ming Dynasty (1892), TCMs are recorded and described in great detail. Most traditional Chinese medicines consist of the caudex, leaf, peel, and radix of plants, which belong to the category of botanical medicines. The remaining medicines are composed of animal and mineral medicines. A valuable TCM named ginseng is shown in Figure 7.1 as an example to give an intuitive impression of botanical medicine in TCMs, whereas the antler of *Cervus nippon* Temminck is given as an instance of animal medicine in Figure 7.2.

Figure 7.1 The radix of ginseng.

Figure 7.2 The *Cervus nippon* Temminck and its antler.

In clinical therapeutics, TCMs are in most cases administered to patients as formulations, which usually consist of over 10 to 20 herbs that collectively exert therapeutic actions and modulating effects. Traditionally defined herbal properties, related to the pharmacodynamic, pharmacokinetic, and toxicological, as well as physicochemical properties of their principal ingredients, have been used as the basis for formulating TCM multiherb prescriptions [1]. TCM formulations can be classified in accordance with several categories, namely the Four Natures, the Five Tastes, and the Channel Tropisms [2]. The Four Natures are related to the degree of Yin and Yang, ranging from cold (extreme Yin), cool, neutral semiwarm, to hot (extreme Yang). The Five Tastes of TCM formulations are pungent, sweet, sour, bitter, and salty.

Recent years have seen increasing interest in the use of traditional Chinese medicine for the prevention and treatment of different kinds of complex diseases, such as cancer, especially when modern Western medicine is not capable of providing a "cure-all" solution for human diseases [3]. The reason why TCM formulations have been successfully applied in clinical practice lies to a large extent in the synergistic effect of different bioactive ingredients [4], which may be simply explained as the fact that the combination of multiple herbs in certain proportion has greater efficacy than the sum of each single herb [5].

Despite successful applications, it should be, however, pointed out that it is really difficult at the present time for us to clearly figure out the therapeutic mechanism of formulations of traditional Chinese medicines. This may be the main factor that prevents most of the TCMs or their prescriptions from going into the Western market. As is known to all, one of the characteristics of Oriental herbal medicine preparations is that all the herbal medicines, either presenting as single herbs or as collections of herbs in composite formulae, are extracted with boiling water [6]. Both the physical and chemical changes during the decoction process are of high complexity. In this case, it is usually very hard for us to know how many chemical components exist in the mixture after decoction, what they are, and how much there is. This may be the main reason why quality control of oriental herbal drugs is more difficult than that of western drugs. As pointed out in "General Guidelines for Methodologies on Research and Evaluation of Traditional Medicines" (World Health Organization, 2000) [7],

> Despite its existence and continued use over many centuries, and its popularity and extensive use during the last decade, traditional medicine has not been officially recognized in most countries. Consequently, education, training and research in this area have not been accorded due attention and

support. The quantity and quality of the safety and efficacy data on traditional medicine are far from sufficient to meet the criteria needed to support its use world-wide. The reasons for the lack of research data are due to not only to health care policies, but also to a lack of adequate or accepted research methodology for evaluating traditional medicine.

Due to the key role of quality control of traditional Chinese medicine in the development of new drugs originating from TCMs, it is described here in some detail. In general, one or two markers or pharmacologically active components in herbs or herbal mixtures are currently employed for evaluating the quality and authenticity of herbal medicines, in the identification of the single herb or HM preparations, and in assessing the quantitative herbal composition of an herbal product. This kind of determination, however, cannot provide a complete picture of a herbal product, because multiple constituents are usually responsible for its therapeutic effects. These multiple constituents may work "synergistically" and could hardly be separated into active parts. Moreover, the chemical constituents in component herbs in the HM products may vary depending on harvest season, plant origin, drying process, and other factors.

It seems to be necessary to determine most of the phytochemical constituents of herbal products in order to ensure the reliability and repeatability of pharmacological and clinical research, to understand their bioactivities and possible side effects of active compounds, and to enhance product quality control [6,8]. Several chromatographic techniques, such as high-performance liquid chromatography (HPLC), gas chromatography (GC), capillary electrophoresis (CE), and thin layer chromatography (TLC), can be applied for this kind of documentation. In this way, the full herbal product could be regarded as the active "compound."

The concept of phytoequivalence was developed in Germany in order to ensure consistency of herbal products [9]. According to this concept, a chemical profile, such as a chromatographic fingerprint, for herbal products should be constructed and compared with the profile of a clinically proven reference product. In 2004, the Chinese State Food and Drug Administration (SFDA) regulated the compositions of liquid injection with HM ingredients using stringent quality procedures such as chemical assay and standardization. Fingerprints of HM liquid injections are compulsorily carried out for this purpose. In addition, among the various experimental techniques, chromatographic methods are highly recommended for uncovering the fingerprints of these products. By definition, a chromatographic fingerprint of a HM is, in practice, a chromatographic pattern of the extract of some common chemical components of pharmacologically active or chemical characteristics. This chromatographic profile should feature the fundamental attributions of "integrity" and "fuzziness" or "sameness" and "differences" so as to chemically represent the HM investigated. It is suggested that with the help of chromatographic fingerprints obtained, the authentication and identification of herbal medicines can be accurately conducted ("integrity") even if the amount or concentration of the chemically characteristic constituents is not exactly the same for different samples of this HM (hence "fuzziness"), or the chromatographic fingerprints could successfully demonstrate both the "sameness" and "differences" between various samples. Thus, we should globally consider multiple constituents in the HM extracts, and not individually consider only one or two marker components for evaluating the quality of the HM products.

However, in any HM and its extracts, there are hundreds of unknown components and many of them are in low amounts. Moreover, there usually exists variability within the same herbal material. Consequently, to obtain reliable chromatographic fingerprints that represent pharmacologically active and chemically characteristic components is not an easy or trivial task. Fortunately, chromatography offers [6] very powerful separation ability, such that the complex chemical components in HM extracts can be separated into many relatively simple subfractions. Furthermore, the recent approaches of applying hyphenated chromatography and spectrometry such as high-performance liquid chromatography–diode array detection (HPLC–DAD), gas chromatography–mass spectroscopy (GC–MS), capillary electrophoresis–diode array detection (CE–DAD), HPLC–MS, and HPLC–NMR, could provide the additional spectral information, which will be very helpful for the qualitative analysis and even for online structural elucidation. With the help of spectral information the hyphenated instruments show greatly improved performance in terms of the elimination of instrumental interference, retention time shift correction, selectivity, chromatographic separation abilities, and measurement precision [10–13]. If hyphenated chromatography is further combined with chemometric approaches, clear pictures might be developed for the obtained chromatographic fingerprints. These excellent properties are the so-called dimension advantages proposed by Booksh and Kowalski. Without any doubt, the chemical fingerprint technique combined with chemometric methodologies will provide a powerful tool for quality control of traditional Chinese medicines or their formulations.

In quality control, the recognition of authentic traditional Chinese medicines is one important aspect. The chemometric methods for similarity analysis and pattern recognition could be employed to distinguish authentic TCMs from fake ones. As is known, fake medicines will always make money. For this reason, there exist some fake TCMs on the market. Therefore, it's of great value to distinguish authentic medicine from fake ones. We focus here on the use of pattern recognition techniques for achieving this aim. In the following sections, the application of support vector machines to identify authentic products of two important TCMs, namely *Pericarpium Citri Reticulatae* (PCR) and *Pericarpium Citri Reticulatae Viride* (PCRV) is discussed in some detail.

7.3 Recognition of authentic PCR and PCRV using SVM

7.3.1 Background

Pericarpium Citri Reticulatae is the dried peel of the mature tangerine and its mutations, collected from September to December. *Pericarpium Citri Reticulatae Viride* is the dried peel or dried immature fruit of the immature tangerine and its mutations, collected from May to August. As traditional Chinese herbs, they have been commonly used for thousands of years because of their advantages of low toxicity and pharmacological activities. PCR has been mostly used as one of the herbal materials to eliminate phlegm and strengthen the spleen, whereas PCRV is always applied to disperse stagnation and strengthen the liver. Moreover, PCR is also used as a kind of flavoring for its delicious taste and special fragrance, especially PCR from Citrus reticulata "chachi" [14].

In addition to macroscopic and microscopic authentication, chemical identification of TCM materials is an important and useful means as it is directly connected with the medicinal functions of TCM. Hesperidin is usually chosen as a marker compound to assess the quality of PCR and PCRV. Pharmacological and clinical studies indicated that it was a bioactive compound with reported activities such as inhibiting the proliferation of human breast cancer cells, central nervous system depressant action, anti-inflammatory properties, antibacterial and allelopathic activities, and so on [15–19]. However, hesperidin not only exists in both PCR and PCRV, but also in other citrus plants [20]. Therefore, chemical identification and discrimination of these two herbal materials by using hesperidin as a marker compound seems to be insufficient. In light of this, a characteristic chromatographic fingerprint is developed for the purpose. Chemical components that we used to discriminate traditional Chinese herbs are secondary metabolized materials of plants. Although these secondary metabolites are similar because of the same plant source, they are unstable and easily affected by circumstances such as weather, soil, and the like. An index should be compiled to evaluate the similarity and dissimilarity of different herbs or different samples of one certain kind of herb from different sources. So, a chemometric parameter, a correlation coefficient, was employed to solve this problem, which has been widely used for quality control of TCM [21–23].

PCR and PCRV are called couple herbs because their plant sources are the same and their chemical fingerprints are quite similar. Chemical discrimination between them is a big problem. In recent decades, many tangerine peels from new tangerine mutations, such as *Citrus unshiu Marc.*, *Citrus poonensis Tanaka*, and the like, are mixed and used as PCR or PCRV. They are called "mixed peels" and qualities are uncertain according to traditional experience. The presence of many "mixed peels" makes quality control of PCR and PCRV more difficult. In this study, the discrimination abilities of the correlation coefficient, principal component analysis (PCA), and partial least squares linear discrimination analysis (PLS-LDA) were compared.

7.3.2 Data description

A summary of 60 samples is given in Table 7.1. The authentic PCR and PCRV samples from Xinhui (*Citrus reticulata "Chachi"*), Zigong (*Citrus reticulata "Dahongpao"*), and Changsha (*Citrus erythrosa Tanaka*) were collected from their original cultivation places. Those samples were authenticated by Professor Pei-shan Xie (Zhuhai Kingman Institute of Herbal Medicine Research, Guangdong, China). The commercial samples were purchased from their selling places including 11 provinces and cities of China. The specimens are preserved in that institute at the present time. As is known, the chromatographic fingerprint is a powerful tool for quality control of traditional Chinese medicines, therefore we measure here the fingerprints of 60 PCR/PCRV samples by using HPLC-DAD, shown in Figure 7.3. The fingerprints are believed to provide global and comprehensive information on the PCR/PCRV samples and hence could be used for discrimination analysis of the data. Details of these data may be found in our previous work [14].

7.3.3 Recognition of authentic PCR and PCRV using whole chromatography

The 60 samples are currently classified into two groups. The first group consists of 39 authentic PCR/PCRV samples, and the remaining 21 samples belong to commercial ones. Principal component analysis (PCA), as an unsupervised multivariate data analysis approach, is employed first to observe the intrinsic data structure. The scores plot resulting from PCA on the mean-centered full spectral data is shown in Figure 7.4. From the plot, it can be found that most of the 39 authentic samples are well separated from the samples. However, PCA cannot be used to predict whether a given new sample is authentic. In this case, it is necessary to build a predictive model based on supervised discriminant analysis.

Table 7.1 A Summary of the Tested Samples

No.	Sample Name	Source	Year of Collection	No.	Sample Name	Source	Year of Collection
1	PCR-A	Xinhui, Guangdong	2002	31	PCR-C	Purchased from Zhejiang	unknown
2	PCR-A	Xinhui, Guangdong	2003	32	PCR-C	Purchased from Hunan	unknown
3	PCR-A	Xinhui, Guangdong	2003	33	PCR-C	Purchased from Henan	unknown
4	PCR-A	Xinhui, Guangdong	2003	34	PCR-C	Purchased from Henan	unknown
5	PCR-A	Xinhui, Guangdong	2003	35	PCR-C	Purchased from Guangdong	unknown
6	PCR-A	Xinhui, Guangdong	2003	36	PCRV-A	Xinhui, Guangdong	2004
7	PCR-A	Xinhui, Guangdong	2003	37	PCRV-A	Xinhui, Guangdong	2004
8	PCR-A	Xinhui, Guangdong	2003	38	PCRV-A	Zigong, Sichuan	2004
9	PCR-A	Xinhui, Guangdong	2003	39	PCRV-A	Zigong, Sichuan	2004
10	PCR-A	Xinhui, Guangdong	2003	40	PCRV-A	Zigong, Sichuan	2004
11	PCR-A	Xinhui, Guangdong	2004	41	PCRV-A	Zigong, Sichuan	2004
12	PCR-A	Xinhui, Guangdong	2004	42	PCRV-A	Zigong, Sichuan	2004
13	PCR-A	Xinhui, Guangdong	2004	43	PCRV-A	Zigong, Sichuan	2004
14	PCR-A	Xinhui, Guangdong	2004	44	PCRV-A	Zigong, Sichuan	2004
15	PCR-A	Zigong, Sichuan	2004	45	PCRV-A	Zigong, Sichuan	2004
16	PCR-A	Zigong, Sichuan	1995	46	PCRV-A	Zigong, Sichuan	2004
17	PCR-A	Zigong, Sichuan	1997	47	PCRV-A	Zigong, Sichuan	2004
18	PCR-A	Zigong, Sichuan	2000	48	PCRV-A	Zigong, Sichuan	2003
19	PCR-A	Zigong, Sichuan	2004	49	PCRV-A	Changsha, Hunan	2004
20	PCR-A	Zigong, Sichuan	2004	50	PCRV-A	Changsha, Hunan	2004
21	PCR-A	Zigong, Sichuan	2004	51	PCRV-A	Changsha, Hunan	2004
22	PCR-A	Changsha, Hunan	2004	52	PCRV-A	Changsha, Hunan	2004

23	PCR-C	Purchased from Anhui	unknown	53	PCRV-C	Purchased from Henan	unknown
24	PCR-C	Purchased from Guangxi	unknown	54	PCRV-C	Purchased from Guangdong	unknown
25	PCR-C	Purchased from Heilongjiang	unknown	55	PCRV-C	Purchased from Fujian	unknown
26	PCR-C	Purchased from Hebei	unknown	56	PCRV-C	Purchased from Guangxi	unknown
27	PCR-C	Purchased from Henan	unknown	57	PCRV-C	Purchased from Henan	unknown
28	PCR-C	Purchased from Hubei	unknown	58	PCRV-C	Purchased from Anhui	unknown
29	PCR-C	Purchased from Shanghai	unknown	59	PCRV-C	Purchased from Guangdong	unknown
30	PCR-C	Purchased from Shanghai	unknown	60	PCRV-C	Purchased from Heilongjiang	unknown

Note: PCR-A and PCRV-A: Authentic *Pericarpium Citri Reticulatae* (PCR) and *Pericarpium Citri Reticulatae Viride* (PCRV) samples collected from original producing areas (PCR-A and PCRV-A from Xinhui (*Citrus reticulata 'Chachi'*), PCR-A and PCRV-A from Zigong (*Citrus reticulata "Dahongpao"*), PCR-A and PCRV-A from Changsha (*Citrus erythrosa Tanaka*)); PCR-C and PCRV-C: Commercial *Pericarpium Citri Reticulatae* and *Pericarpium Citri Reticulatae Viride* samples purchased from 11 difference provinces and cities.

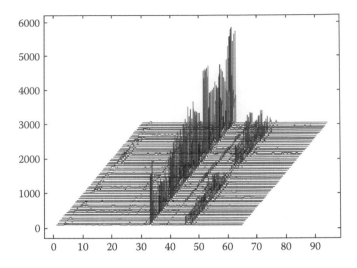

Figure 7.3 The HPLC fingerprint of 60 samples: 39 authentic PCR/PCRV samples and 21 commercial ones.

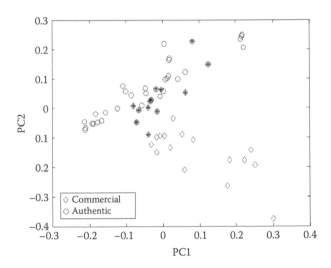

Figure 7.4 The scores plot of the 60 PCR/PCRV samples by using principal component analysis. The points filled with marker "*" denote the support vectors resulting from SVM computation.

To the best of our knowledge, PLS-LDA is one of the most commonly used methods for discriminant analysis [24–26]. Here, it is first utilized to build a predictive model using the full spectra as input. Before analysis, the data are mean-centered. The optimal number of latent variables is chosen by means of ten fold cross-validation. The cross-validation results using

Table 7.2 The Prediction Performance (%) of Ten Fold
Cross-Validation Using PLS-LDA and SVM[a]

Method	Sensitivity	Specificity	Accuracy
PLS-LDA	97.4	81.0	91.7
SVM	97.4	81.0	91.7
SVM+RFE	97.4	100.0	98.3
SVM+SFS	98.3	100.0	95.1
SVM+MIA	100.0	100.0	100.0

[a] The numbers of variables selected by recursive feature elimination (RFE), SFS-based method, and margin influence analysis (MIA) are 7, 35, and 100, respectively.

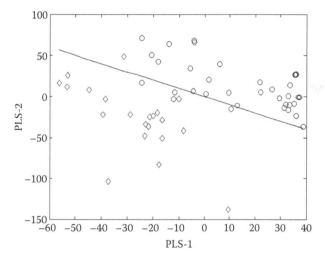

Figure 7.5 The PLS-LDA decision line in the latent variable space.

PLS-LDA are shown in Table 7.2. The sensitivity, specificity, and overall accuracy are 97.4, 81.0, and 91.7%, respectively, which indicates that PLS-LDA can distinguish the two groups of samples with high accuracy. So, it may be deemed a powerful tool for quality control of traditional Chinese medicines. In addition, we also use PLS-LDA to build a classification model using two latent variables. The discrimination line in the latent variable space together with projected samples is shown in Figure 7.5. The PLS-LDA model could be used to predict the class of a future sample if the corresponding HPLC-DAD under the same experimental condition were measured.

Furthermore, support vector machines were used to build a model for recognition of the authentic samples from the commercial ones. The LIBSVM package was used here for all SVM-related calculations [27]. Due to the fact that the data are well separated, we here chose a linear kernel for building the SVM classifier. The used SVM-type is C-SVM, where C is the penalizing factor. We use ten fold cross-validation to choose an optimal value of C, with C belonging to [10, 50, 100, 200, 500]. This procedure resulted in the optimal C = 10. The ten fold cross-validation results for C = 10 are computed using SVM and recorded in Table 7.2. By comparison, PLS-LDA and SVM have the same performance.

As introduced in Chapter 2, the decision function of SVM is only determined on the so-called support vectors (SVs). Loosely speaking, support vectors are the samples located on the "middle band" between groups and usually occupying a small part of the whole sample. To illustrate this point, a SVM classifier is built with the C set to 10 using all 60 samples as input. The resulting SVM model is involved with 13 SVs. The ratio of SVs is 13/60. The 13 PCR/PCRV samples are marked in Figure 7.5 as points with a filled asterisk "*". Obviously, one can find that the 13 SVs are, as expected, positioned in the area between the two classes of samples. The remaining 47 samples have no influence on the SVM model.

In addition, the coefficient associated with each variable resulting from both PLS-LDA and SVM is plotted in the upper and bottom panel of Figure 7.6, respectively, for visual comparison. The middle panel shows

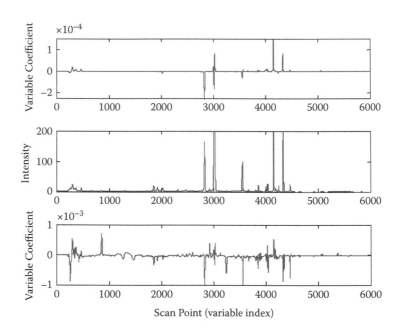

Figure 7.6 The middle plot shows the mean chromatogram of the 60 samples. The variable coefficients of the PLS-LDA model are plotted in the upper panel, and the bottom panel presents the variable coefficients of the support vector classification model.

the mean HPLC-DAD fingerprint of all 60 samples. By comparison, it appears that both PLS-LDA and SVM could have the potential to identify the chromatography peaks.

7.3.4   *Variable selection improves performance of SVM*

One of the greatest characteristics of the data resulting from modern analytical instruments is that the number of variables is much larger than that of samples. This is the so-called "large p, small n" problem. An important aspect in analyzing such data is variable selection. Concerning support vector machines, it has been demonstrated that variable selection could benefit the predictive performance and the interpretability of a SVM model [11, 28–33]. For instance, Guyon et al. utilized a recursive feature elimination (RFE) [29] strategy to rank the variable according to the weight value of the SVM classifier and greatly improved the performance of SVM. Gualdrón et al. recently proposed a method for variable selection for SVM by analyzing the absolute margin changes after eliminating a variable [28] and also achieved better predictive ability compared to that using all variables. Based on model population analysis (MPA) [34,35], Liang and Li et al. recently [30] proposed a variable selection method, called margin influence analysis (MIA), by investigating the influence of variables on the margin of the support vector machine model. To investigate whether variable selection would improve the performance of the SVM, RFE and MIA were employed in this study to analyze the PCR/PCRV data.

We first use RFE for variable selection of the SVM model. In the original RFE algorithm, the variables are eliminated one by one. However, it should be noted that the computational cost would be very high if the data contain a large number of variables (in these data, 6,000 variables). To reduce cost, the authors also stated that more than one variable could be removed each time [29]. Here, we propose to use the exponentially decreasing function (EDF) to control the process of backward elimination of variables. The details of EDF can be found in previous work on variable selection of near-infrared data using competitive adaptive reweighted sampling (CARS) [36,37].

An example of EDF is shown in Figure 7.7. As can be clearly seen, the process of variable reduction can be roughly divided into two stages. In the first stage, variables are eliminated rapidly which performs a "fast selection," whereas in the second stage, variables are reduced in a very gentle manner, which is called instead a "refined selection" stage in our study. Therefore, variables with little or no information in a full spectrum can be removed in a stepwise and efficient way because of the advantage of EDF.

With the introduced EDF, recursive feature selection for support vector machines was performed on our PCR/PCRV data. The number

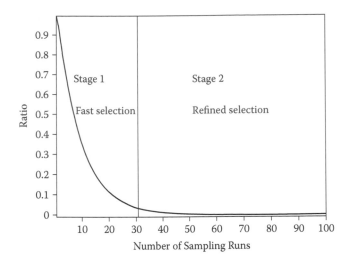

Figure 7.7 Illustration of exponentially decreasing function (EDF).

of iterations was set to 50. Before running RFE+SVM, each variable was standardized to range from 0 to 1. We chose the squared weight associated with the variable as the index for variable importance evaluation. At each iteration, ten fold cross-validation (CV) was utilized to evaluate the prediction ability of SVMs using the corresponding variable subset.

Figure 7.8 presents the results of RFE+SVM. The upper panel shows the number of variables (nVar) at each iteration, which is a realization of the proposed exponentially decreasing function. Shown in the middle panel is the weight of the variable in the SVM model at each iteration during the RFE procedure. Note that each line in this plot denotes the weight changes for a variable. Correspondingly, the squared weight (variable importance) of each variable at each iteration is shown at the bottom. By using RFE+SVM with the iteration number set to 50, we finally obtained 50 subsets of variables. For each subset, ten fold CV was taken to estimate the prediction ability. The subset at the 42nd iteration consisting of 7 variables achieved the best performance in terms of ten fold CV prediction error. The sensitivity, specificity, and overall accuracy are presented in Table 7.2. Compared to the results obtained from the full spectrum model, the results after variable selection were improved.

For comparison, the sequential forward selection (SFS)-based method of variable selection for support vector machines was also investigated [28]. The linear kernel function was chosen and the penalizing factor C optimized by ten fold cross-validation. According to the literature [28], the squared norm of the weight vector of the optimal separation hyperplane (OSH) when all variables except for the ith one were used served as the index for evaluating the importance of the ith variable. For the

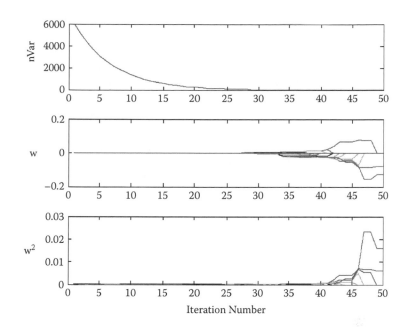

Figure 7.8 The upper panel shows the number of variables (nVar) at each iteration. The weight and squared weight of each variable at each iteration of recursive feature elimination combined with SVM are shown in the middle and bottom panels, respectively.

PCR/PCRV data of 6,000 scan points, a SVM model was built by using all 6,000 variables and the squared norm of the OSH weight vector was recorded. Then in an iterative procedure, we computed 6,000 SVM models, each of which contained 5,999 variables by removing only one variable. The importance of all the variables was calculated and is shown in Figure 7.9.

Then all the variables were ranked by their importance level. Based on the SFS strategy, variables were added to the SVM classifier starting from the one of the highest importance. The other variables were included in the model one by one if they improved the model's predictive performance. This process stopped when adding a new variable did not improve cross-validated results. In this way, 35 variables were finally selected. The ten fold cross-validated results in terms of sensitivity, specificity, and accuracy are presented in Table 7.2. Compared to SVM+RFE, the results of the SFS-based method are comparable to those from SVM+RFE.

Furthermore, the newly proposed margin influence analysis technique is also taken for variable selection of SVMs. To begin with, the MIA algorithm is first described [30]. It works mainly in three successive steps: (1) obtain N subdatasets by Monte Carlo sampling (MCS) in the variable

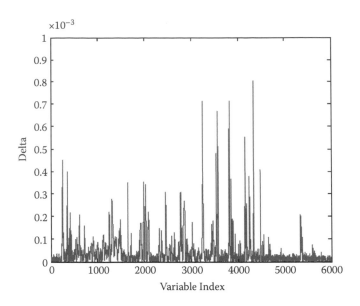

Figure 7.9 The variable importance computed using the SFS-based method.

space; (2) establish a submodel for each subdataset; and (3) statistically analyze the margins of all the submodels. Details of MIA are presented below.

First suppose that we are given a dataset $(\mathbf{X}, \mathbf{y})$ consisting of m samples
in the rows and p variables in the columns. The class label vector $\mathbf{y}$ is of
size m × 1, with element equal to 1 or –1 for the binary classification case.
The number of MCS is set to N (usually large, e.g., 10,000). With such a setting, Monte Carlo sampling in the variable space can be conducted in three
steps: (1) predefine the number of variables, denoted by Q, to be sampled;
(2) in each sampling, Q variables are randomly picked from all p variables
without replacement and thus a subdataset of size m × Q is obtained; repeat
this procedure N times, and finally get N subdatasets. All the sampled N
subdatasets are denoted $(\mathbf{X}_{sub}, \mathbf{y}_{sub})_i$, i = 1, 2, 3, ... , N. Figure 7.10 shows the
sampling scheme in the variable space. Taking the first sampling as an
example, only a few variables marked by black squares are selected.

Second, for each of the randomly sampled subdatasets, one can build
a linear kernel-based SVM classifier given a penalizing factor C. In the
current work, C is chosen by cross-validation. Therefore, N SVM classifiers together with N margins can be computed. The N margins are denoted
by M_i, i = 1, 2, ... , N.

The third step is to perform statistical analysis of the margins of the SVM
models associated with each variable, aiming at uncovering informative variables. Without loss of generality, we take the *i*th variable as a case to illustrate
the computing procedure. Assume that all N computed SVM classifiers are

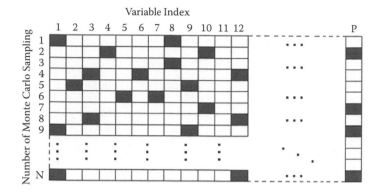

Figure 7.10 Scheme of Monte Carlo sampling in the variable space. In each sampling, only a few variables are randomly chosen, denoted by black squares, to build a model.

classified into two groups, namely Group A and Group B. Group A collects all the models including the ith variable, and Group B collects all the models not including the ith variable. Assume that the numbers of the models in Group A and B are $N_{i,A}$ and $N_{i,B}$, respectively. Then we have

$$N_{i,A} + N_{i,B} = N \tag{7.1}$$

Naturally, we can also get $N_{i,A}$ and $N_{i,B}$ margins associated with SVM classifiers in Group A and Group B, respectively. Furthermore, we can compute two distributions corresponding to the $N_{i,A}$ and $N_{i,B}$ margins, respectively. Denote the mean values of the two distributions by $MEAN_{i,A}$ and $MEAN_{i,B}$, respectively. The difference of the two mean values can therefore be calculated as

$$DMEAN_i = MEAN_{i,A} - MEAN_{i,B} \tag{7.2}$$

From Equation 7.2, one can expect that the inclusion of the ith variable in a model can increase the margin of a SVM model if $DMEAN_i > 0$. This type of variable is treated as the candidate of informative variables in the present study. On the contrary, if $DMEAN_i < 0$, one may infer that including this variable in a model will decrease the margin of the SVM models and thus reduce the predictive performance of a SVM model. By analogy, this type of variable is called an uninformative variable. The two kinds of variables are illustrated in Plots A and B of Figure 7.11, respectively.

After deriving the margin's distribution of each variable, we proceed to identify the informative variables in three successive steps: (1) remove all the variables with $DMEAN_i < 0$, (2) use the Mann–Whitney U-test to check whether the increment of margin is significant, leading to a p-value

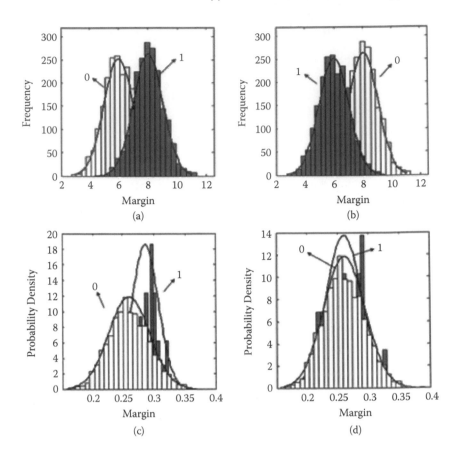

Figure 7.11 The expected paired distributions of the informative variable as well as an uninformative one are shown in Plots A and B, respectively. Illustration of an informative variable (the 4,150th variable) and an uninformative one (the 934th variable) for the TCM data are shown in Plots C and D, respectively.

for each variable, and (3) rank the variables using the p-value. In this sense, the variables with p-value smaller than a predefined threshold (e.g., 0.05) are defined as informative variables. The informative variables should be treated as the most possible biomarker candidates. It should be noted that the proposed MIA method can also be applied to SVM with non-linear kernels, such as Gaussian kernels because only the distribution of margins is required. By the way, the reason for using the Mann–Whitney U-test rather than the two-sample t-test is that the Mann–Whitney U-test is a nonparametric test method without any distribution-related assumptions and is also a robust method.

Before analysis using MIA, the data are range-scaled into the region [0, 1]. By running MIA + SVM, all 6,000 variables are ranked according to

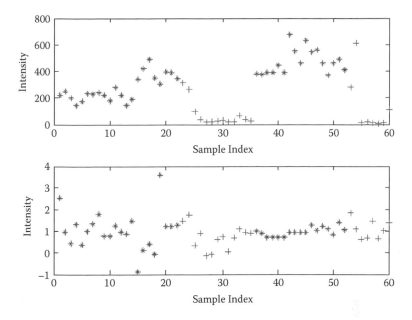

Figure 7.12 The intensity of one informative variable (upper panel) as well as one uninformative one (bottom panel) identified by margin influence analysis.

their significance level. Then by using ten fold cross-validation, the first most significant 100 variables are chosen as an optimal subset. For illustration, an informative variable (the 4,150th variable, with separated distribution) together with an uninformative one (the 934th variable, with heavily overlapping distribution) identified by MIA is shown in Figures 11C and D, respectively. To examine the power of MIA, the intensities across different samples of the informative variable and the uninformative one are plotted in Figure 7.12. Apparently, the intensity distribution of the informative variable has clear separation between the two classes of samples. By contrast, no significant difference exists concerning the intensity distribution of the uninformative one in either class of samples.

The 100 variables are denoted in Figure 7.13. It's very interesting to note that the selected 100 variables are nearly all located at the chromatographic peaks, suggesting that MIA is a very good alternative for identifying the informative variables. The sensitivity, specificity, and overall accuracy are all 100%, shown in Table 7.2. In addition, the selected 100 variables are also used for building the PLS-LDA model. The sensitivity, specificity, and accuracy of cross-validated PLS-LDA are 97.4, 100.0, and 98.3%, indicating the variables selected by MIA+SVM can also achieve good performance for PLS-LDA. PCA is also performed on the reduced data after range-scaling. The scores plot is shown in Figure 7.14. Clearly, the authentic samples are well separated from the commercial ones. From

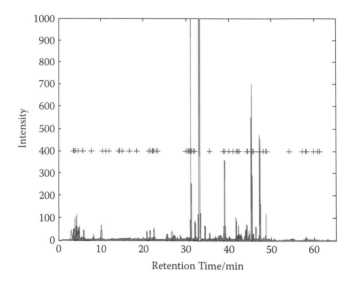

Figure 7.13 Selected variables by margin influence analysis (MIA) are denoted by the plus "+".

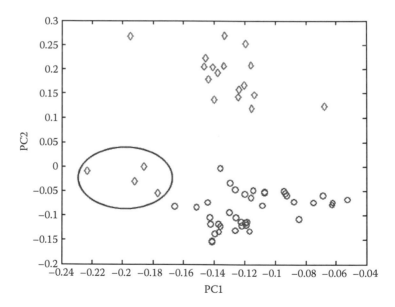

Figure 7.14 PCA scores plot using the first 100 significant variables using MIA. The four commercial samples encompassed by the ellipse may be deemed the authentic ones.

this plot, it can be found that the four commercial samples (encompassed by the ellipse) are located near the authentic ones. In this sense, they may be deemed the authentic ones, which is consistent with results reported in our previous work [14].

7.4 Some remarks

Traditional Chinese medicines represent a kind of complex analytical system. The qualitative and quantitative analysis of TCMs and their prescriptions are highly dependent on modern analytical instruments such as gas chromatography–mass spectroscopy (GC–MS) and liquid chromatography–mass spectroscopy (LC–MS). There is strong evidence showing that the data analysis of traditional Chinese medicines is becoming more and more important due to the huge amount of data resulting from high-throughput instruments. In this context, the statistical and chemometric methods are beginning to play a more and more important role in the research and development of traditional Chinese medicines. Support vector machines have found some applications in the field of TCMs [1,38]. We hope that more and more effort and research will be dedicated to SVM applications to TCMs.

References

1. Wang, J.F., Cai, C.Z., Kong, C.Y., Cao, Z.W., and Chen, Y.Z. 2005. A computer method for validating traditional Chinese medicine herbal prescriptions. *Amer. J. Chinese Med.,* 33(2):281–297.
2. Wang, X., Qu, H., Liu, P., and Cheng, Y.A. 2004. Self-learning expert system for diagnosis in traditional Chinese medicine. *Expert Syst. Appl.,* 26: 557–566.
3. Normile, D. 2003. Asian medicine. The new face of traditional Chinese medicine. *Science,* 299:188–190.
4. Xue, T. and Roy, R. 2003. Studying traditional Chinese medicine. *Science,* 300:740–741.
5. Oka, H., Yamamoto, S., Kuroki, T., Harihara, S., Marumo, T., Kim, S.R., Monna, T., Kobayashi, K., and Tango, T. 1995. Prospective study of chemoprevention of hepatocellular carcinoma with Sho-saikoto(TJ-9). *Cancer Cell,* 76:743–749.
6. Liang, Y.-Z., Xie, P.-S., and Chan, K. 2004. Quality control of herbal medicines. *J. Chromatograph. B,* 812:53–70.
7. WHO 2000. General Guidelines for Methodologies on Research and Evaluation of Traditional Medicines, page 1.
8. Chau, F.-T., Chan, H.-Y., Cheung, C.-Y., Xu, C.-J., Liang, Y., and Kvalheim, O.M. 2009. Recipe for uncovering the bioactive components in herbal medicine. *Anal. Chem.,* 81(17):7217–7225.
9. Tyler, V.E. 1999. Phytomedicines: Back to the future. *J. Nat. Prod.,* 62(11):1589–1592.

10. Li, B.Y., Liang, Y.Z., Hu, Y., Xie, P.S., Yu, R.Q., Li, F.M., and Sun, Y. 2004. Correction of retention time shift of components for chromatographic fingerprints of herbal medicine. *Chinese J. Anal. Chem.*, 32(3):313–316.

11. Yi, L.Z., Yuan, D.L., Liang, Y.Z., Xie, P.S., and Zhao, Y. 2009. Fingerprinting alterations of secondary metabolites of tangerine peels during growth by HPLC-DAD and chemometric methods. *Anal. Chim. Acta*, 649(1):43–51.

12. Li, B.Y., Hu, Y., Liang, Y.Z., Xie, P.S., and Ozaki, Y. 2006. Modified secured principal component regression for detection of unexpected chromatographic features in herbal fingerprints. *Analyst*, 131(4):538–546.

13. Li, B.Y., Hu, Y., Liang, Y.Z., Xie, P.S., and Du, Y.P. 2004. Quality evaluation of fingerprints of herbal medicine with chromatographic data. *Anal. Chim. Acta*, 514(1):69–77.

14. Yi, L.-Z., Yuan, D.-L., Liang, Y.-Z., Xie, P.-S., and Zhao, Y. 2007. Quality control and discrimination of *Pericarpium Citri Reticulatae* and *Pericarpium Citri Reticulatae Viride* based on high-performance liquid chromatographic fingerprints and multivariate statistical analysis. *Anal. Chim. Acta*, 588(2):207–215.

15. Sheu, F., Chuang, W.I., and Chien, P.J. 2007. *Citri Reticulatae Pericarpium* extract suppresses adipogenesis in 3T3-L1 preadipocytes. *J. Sci. Food Agric.*, 87:2382–2389.

16. Wang, D.D., Wang, J., Huang, X.H., Tu, Y., and Ni, K.Y. 2007. Identification of polymethoxylated flavones from green tangerine peel (*Pericarpium Citri Reticulatae Viride*) by chromatographic and spectroscopic techniques. *J. Pharmaceut. Biomed. Anal.*, 44(1):63–69.

17. Shi, Q., Liu, Z., Yang, Y., Geng, P., Zhu, Y.Y., Zhang, Q., Bai, F., and Bai, G. 2009. Identification of anti-asthmatic compounds in *Pericarpium Citri Reticulatae* and evaluation of their synergistic effects. *Acta Pharmacol. Sinica*, 30(5):567–575.

18. Wang, Y.M., Yi, L.Z., Liang, Y.Z., Li, H.D., Yuan, D.L., Gao, H.Y., and Zeng, M.M. 2008. Comparative analysis of essential oil components in *Pericarpium Citri Reticulatae Viride* and *Pericarpium Citri Reticulatae* by GC-MS combined with chemometric resolution method. *J. Pharmaceut. Biomed. Anal.*, 46(1):66–74.

19. Zhou, X., Sun, S.-Q., and Huang, Q.-H. 2008. Identification of *Pericarpium Citri Reticulatae* from different years using FTIR. *Guang Pu Xue Yu Guang Pu Fen Xi*, 28(1):72–74.

20. Peng, Y., Liu, F., and Ye, J. 2006. Quantitative and qualitative analysis of flavonoid markers in *Frucus aurantii* of different geographical origin by capillary electrophoresis with electrochemical detection. *J. Chromatog. B*, 830(2):224–230.

21. Fang, K.T., Liang, Y.Z., Yin, X.L., Chan, K., and Lu, G.H. 2006. Critical value determination on similarity of fingerprints. *Chemometr. Intell. Lab. Syst.*, 82(1–2):236–240.

22. Lu, G.H., Chan, K., Liang, Y.Z., Leung, K., Chan, C.L., Jiang, Z.H., and Zhao, Z.Z. 2005. Development of high-performance liquid chromatographic fingerprints for distinguishing Chinese Angelica from related umbelliferae herbs. *J. Chromatog. A*, 1073(1–2):383–392.

23. Lu, H.M., Liang, Y.Z., and Chen, S. 2006. Identification and quality assessment of Houttuynia cordata injection using GC-MS fingerprint: A standardization approach. *J. Ethnopharmacol.*, 105(3):436–440.

24. Barker, M. and Rayens, W. 2003. Partial least squares for discrimination. *J Chemometr*, 17:166–173.
25. Geladi, P. and Kowalski, B.R. 1986. Partial least-squares regression: A tutorial. *Anal. Chim. Acta*, 185:1–17.
26. Wold, S., Sjöström, M., and Eriksson, L. 2001. PLS-regression: A basic tool of chemometrics. *Chemometr. Intell. Lab.*, 58(2):109–130.
27. http://www.csie.ntu.edu.tw/~cjlin/libsvm.
28. Gualdrón, O., Brezmes, J., Llobet, E., Amari, A., Vilanova, X., Bouchikhi, B., and Correig, X. 2007. Variable selection for support vector machine based multisensor systems. *Sens. Actuat. B: Chem.*, 122(1):259–268.
29. Guyon, I., Weston, J., Barnhill, S., and Vapnik, V. 2002. Gene selection for cancer classification using support vector machines. *Mach. Learn.*, 46(1):389–422.
30. Li, H.-D., Liang, Y.-Z., Xu, Q.-S., Cao, D.-S., Tan, B.-B., Deng, B.-C., and Lin, C.-C. 2010. Recipe for uncovering predictive genes using support vector machines based on model population analysis (accepted). *IEEE/ACM Trans. Comput. Biol. Bioinf.*
31. Bierman, S. and Steel, S. 2009. Variable selection for support vector machines. *Commun. Statist. Simul. Comput.*, 38(8):1640–1658.
32. Zhang, H.H. 2006. Variable selection for support vector machines via smoothing spline ANOVA. *Statist. Sinica*, 16(2):659–674.
33. Louw, N. and Steel, S.J. 2006.Variable selection in kernel Fisher discriminant analysis by means of recursive feature elimination. *Comput. Statist. Data Anal.*, 51(3):2043–2055.
34. Li, H.-D., Liang, Y.-Z., Xu, Q.-S., and Cao, D.-S. 2010. Model population analysis for variable selection. *J. Chemometr.*, 24(7–8):418–423.
35. Li, H.-D., Zeng, M.-M., Tan, B.-B., Liang, Y.-Z., Xu, Q.-S., and Cao, D.-S. 2010. Recipe for revealing informative metabolites based on model population analysis. *Metabolomics*, 6(3):353–361.
36. Li, H.-D., Liang, Y.-Z., Xu, Q.-S., and Cao, D.-S. 2009. Key wavelengths screening using competitive adaptive reweighted sampling method for multivariate calibration. *Anal. Chim. Acta*, 648(1):77–84.
37. http://code.google.com/p/carspls/.
38. Wang, Y., Wang, X.-W., and Cheng, Y.-Y. 2006. A computational approach to botanical drug design by modeling quantitative composition–activity relationship. *Chem. Biol. Drug Des.*, 68:166–172.

chapter eight

Support vector machines applied to OMICS study

Contents

8.1 Introduction

In the past decade, there has been growing interest in OMICS study which informally refers to a field of study in biology ending in -omics, such as genomics, proteomics, metabolomics, and so on. To the best of our knowledge, the OMICS study is central to systems biology in life science. In this field, an overwhelming amount of data is produced daily and data analysis using statistical learning algorithms is the key for mining useful information and knowledge of the complex biological systems. In this chapter, we just focus on the use of support vector machines in exploring the biological data produced in the OMICS study.

8.2 A brief description of OMICS study

A number of research fields on OMICS have emerged, such as genomics, transcriptomics, proteomics, metabolomics, lipidomics, and so on. Here, an overview of genomics, proteomics, and metabolomics is given, aiming at providing a global picture of what OMICS means.

Genomics is the study of the whole genomes of organisms. The field includes intensive efforts to determine the entire DNA sequences of organisms and fine-scale genetic mapping efforts. It also includes studies of intragenomic phenomena such as heterosis, epistasis, pleiotropy, and other interactions between loci and alleles within the genome. In contrast, the investigation of the roles and functions of single genes is a primary focus of molecular biology or genetics and is a common topic of modern medical and biological research. Research of single genes does not fall into the definition of genomics unless the aim of this genetic, pathway, and functional information analysis is to elucidate its effect on, place in, and response to the entire genome's networks. According to the United States Environmental Protection Agency, "the term 'genomics' encompasses a broader scope of scientific inquiry associated technologies than when genomics was initially considered. A genome is the sum total of all an individual organism's genes. Thus, genomics is the study of all the genes of a cell, or tissue, at the DNA (genotype), mRNA (transcriptome), or protein (proteome) levels" [1].

Proteomics is the large-scale study of proteins, particularly their structures and functions. The word "proteome" is a blend of "protein" and "genome," and was coined by Marc Wilkins in 1994 while working on the concept as a PhD student. The proteome is the entire complement of proteins, including the modifications made to a particular set of proteins produced by an organism or system. This will vary with time and distinct requirements, or stresses, that a cell or organism undergoes. The final goal of proteomics is to identify simultaneously the whole proteome of a particular cell, an organelle or tissue type [2]. Although significant progress has been achieved by now [3,4], there still exist great challenging problems in proteomics because the proteome is always dynamic [5,6] and the biological sample is usually of high complexity. However, the qualitative and quantitative changes of the proteome can reveal the dynamic behavior of the physiological function of a species at the molecular level [7] and can further serve as a detecting window into the physiological state. Hence proteomics is now deemed the most powerful tool for clinical research, disease diagnostics, and drug discovery among other disciplines.

Metabolomics is the "systematic study of the unique chemical fingerprints that specific cellular processes leave behind," specifically, the study of their small-molecule metabolite profiles [8]. The metabolome represents the collection of all metabolites in a biological cell, tissue, organ, or organism that are the endproducts of cellular processes [9]. Thus, although mRNA gene expression data and proteomic analyses do not tell the whole story of what might be happening in a cell, metabolic profiling can give an instantaneous snapshot of the physiology of that cell. In metabolomics, the high-throughput analysis of metabolites is mainly performed by two kinds of instruments: nuclear magnetic resonance (NMR) [10–14] and mass spectroscopy (MS) [15–19].

8.3 Support vector machines in genomics

The developed microarray allows scientists to monitor expression levels of thousands of genes associated with different diseases in a very quick and efficient manner. In combination with bioinformatics data analysis methods, such technologies have been gaining extensive applications in the field of cancer classification, aiming at first uncovering the genetic causes that underlie the development of many kinds of human disease [20–24] and then administering an appropriate therapy to the patients. However, the number of genes resulting from microarray experiments is in most cases very large. On the contrary, the number of tissue samples is very small. Moreover, the disease-relevant genes usually occupy only a small percent, making it difficult to identify the few informative genes from the large pool of candidates. However, from the point of view of clinical practice, it is of great value to identify the small number of informative genes for a thorough understanding of the pathogenesis and accurate prediction of clinical outcome. So, many variable selection methods have been proposed or applied to seek the potential genes responsible for tissue phenotypes, for example, class distinction correlation [21], recursive feature elimination [25], entropy method [26], lasso [27], and so on.

To identify the informative genes from the large pool of candidates, based on the newly proposed framework of model population analysis (MPA) [18,28], we have recently developed a new method, called margin influence analysis (MIA) [29], which is specially designed for variable selection of support vector machines. Briefly speaking, the proposed MIA method works by first computing a large number of SVM classifiers using randomly sampled variables. Each sub-model is associated with a margin. Then the nonparametric Mann–Whitney U-test [30] is employed to calculate a p-value for each variable, aiming at statistically identifying the variable that can significantly increase the SVM margin. The rationale behind MIA is that the generalization performance of SVM is heavily dependent on the classifier margin. As is known, the larger the margin is, the better the prediction performance will be. Thus, the variables that can increase the margin of SVM classifiers should be regarded as informative variables or biomarkers. Details of MIA may also be found in Chapter 7. Here, benchmark gene expression data on the colon are used as an example to show the necessity of variable selection.

The original colon dataset contains the expression profiles of 6,500 human genes measured on 40 tumor and 22 normal colon tissues by applying the Affymetrix gene chip technology. A subset of 2,000 genes with the highest minimal intensity across the samples has been screened out by Alon et al. [24] and has also been made publicly available at http://microarray.princeton.edu/oncology/. The heat map of these data is shown in Figure 8.1.

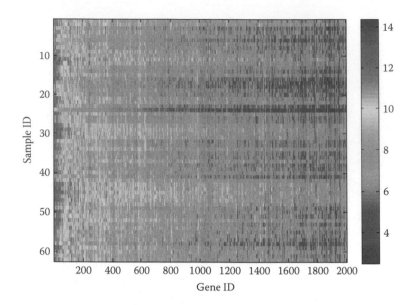

Figure 8.1 The heat map of colon data by using logarithm transformed gene expression intensity with base 2.

To make the classification model easy to explain, we chose a linear kernel for the SVM. There are a total of three tuning parameters for the MIA algorithm [29], that is, C: penalizing factor of SVM, Q: number of sampled variables for drawing subdataset, and N: the number of Monte Carlo sampling. For these data, C is chosen by cross-validation for each subdataset. Q is set to 200 after the performances of different Q values are compared. For Monte Carlo sampling, the larger N is, the better the results will be (but at higher computational cost). Considering the computational cost, N in the present work is set to 10,000. Before running MIA, each gene is standardized to have zero mean and unit variance across all the samples. Considering the fact that the number of samples is small leave-one-out cross-validation (LOOCV) based misclassification error is employed to assess the performance of the selected genes following the references [31,32] and so on.

Using the MIA, 1,219 out of the 2,000 genes are identified as uninformative genes that decrease the margin of SVM classifiers. After removing these genes, the nonparametric Mann–Whitney U-test is applied to test whether the remained 781 genes can significantly increase the margin of the SVM classifiers, leading to a p-value associated with each gene. In all, 217 out of the 781 genes are found to be informative with $p \leq 0.05$. Here, the margin distribution of an informative gene together with an uninformative one is shown in plot A (Gene ID = 1,482, $p = 5.64 \times 10^{-181}$) and plot B (Gene ID = 1,781) in Figure 8.2, respectively. It is clear that the margin

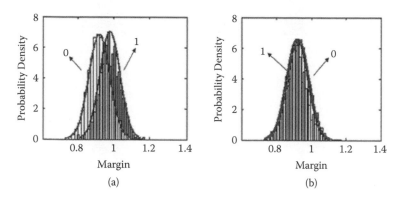

Figure 8.2 The paired distributions of the margins of an informative gene as well as an uninformative one.

distribution when including the 1,482nd gene is right-shifted (toward the larger direction of the margin). This means that this gene has the potential to increase the margin of SVM classifiers and can hence improve the generalization performance if included into a SVM model. In contrast, the 1,781st gene decreases the margin and should therefore be removed from the model. In this sense, it is an uninformative gene. The above analysis indicates that informative genes can be statistically identified by testing the difference of the interesting parameter (i.e., margin) for SVM when a gene is included and excluded in a model.

To build a classification model for cancer prediction, a subset of genes should first be identified. Here, we first rank the genes (DMEAN > 0) using the p-value. For colon data, nine different gene sets are investigated here. The numbers of the nine gene sets are 10, 25, 50, 75, 100, 200, 500, 1,000, and 2,000, respectively. Note that the prediction errors of the established SVM classifiers could not be exactly reproduced due to the embedded Monte Carlo strategy of MIA. Therefore the MIA procedure was run 20 times on the colon data. The mean LOOCV errors as well as the standard deviations are shown in Figure 8.3. From this plot, it can be found that both the mean LOOCV errors and the standard deviation first gradually decrease and then achieve the minimum when 100 significant genes are included. For these data, the results after gene selection are greatly improved compared to those using all genes, indicating that gene selection is very necessary for improving prediction ability, and the identified informative genes by MIA are actually predictive.

By comparison, the results from MIA are very competitive with those reported in the literature. The minimal classification error from Dettling and Buhlmann [33] was 14.52% by using LogitBoost. In Nguyen and Rocke's work [32], the lowest error achieved was 6.45% by using PLS-LD. Sigmoid maximum rank correlation (SMRC) was utilized by Huang and

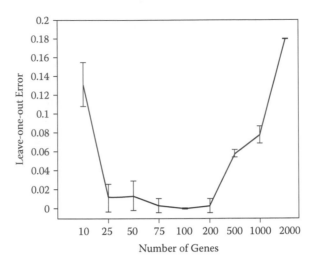

Figure 8.3 The mean LOOCV prediction errors as well as the standard deviation at different numbers of genes.

Table 8.1 The Results on Colon Data by Using RFE, SFS-Motivated Method and MIA

Number of Genes	10	25	50	75	100	200	500	1,000	2,000
RFE	0.00	3.23	0.00	3.23	4.84	4.84	11.29	9.68	17.74
SFS-motivated method	16.13	1.61	0.00	0.00	0.00	0.00	4.84	6.45	17.74
MIA	13.06	1.05	1.29	0.24	0.00	0.24	6.13	7.90	17.74

Ma, leading to the mean classification error 14% with a standard deviation 7%. By using support vector machines, Furey et al. [31] misclassified six samples, resulting in a LOOCV error 9.68%. Besides, we have also encoded two variable selection methods: recursive feature selection (RFE) [25] and sequential forward selection (SFS)-motivated method [34] and performed variable selection on the colon data. The results are presented in Table 8.1. By comparison, it could be found that the proposed MIA is very competitive in gene selection for predicting colon cancer.

Summing up, it might be concluded that MIA is a good alternative for gene selection and the MIA-based SVM. By using MIA, one can distinguish the informative variables from the uninformative genes in an easy and elegant manner. It's expected that MIA will find more applications in other fields, such as proteomics and metabolomics.

8.4   Support vector machines for identifying proteotypic peptides in proteomics

The final goal of proteomics is trying to identify simultaneously the whole proteome of a particular cell, organelle, or tissue type [2]. The identification of proteins is mainly based on real-time mass spectrometry [7,35]. Due to the high complexity of biological and clinical samples, protein mixtures are usually first digested with trypsin. The resulting peptide mixtures are then partially purified by chromatography and sequentially analyzed by a tandem mass spectrometer (MS/MS) [36,37]. The acquired tandem mass spectra are searched against the database using the developed software (e.g., SEQUEST [37], Mascot, SpectraST) to find the best matched peptide of each tandem mass spectrum and further identify the proteins present in the prepared sample. The ultimate goal of basic proteomics is to qualitatively and quantitatively analyze the proteins present in the sample before digestion, rather than the peptides resulting from digested proteins. In the protein-level analysis, the false positive rate (FPR) is in most cases very high, begging the question of how we can control and estimate the FPR effectively.

Recently, Mallick et al. proposed an alternative strategy with the aid of the experimentally discovered "proteotypic" peptides that can uniquely identify a protein [38]. In other words, proteotypic peptides are the peptides that can be detected repeatedly and consistently with high confidence for a given mass spectrometer platform, such as MALDI and ESI, and can provide valuable information for the identification of proteins in biological and clinical research. They also confirmed that the observability of such peptides depends on the experimental platform. For instance, the experimentally determined proteotypic peptides for MALDI and ESI were not synonymous. They also claimed that the possible applications of proteotypic peptides include the validation of protein identification, absolute quantification of proteins, annotation of coding sequences in genomes, and the like.

It has been argued that proteomics are currently turning from the discovery phase into the scoring phase [39]. In the scoring phase, proteotypic peptides would be of great importance. But at present the proteotypic peptides are only identified by running multiple replicate experiments. So it's obvious that identifying the proteotypic peptides is quite a costly and time-consuming procedure. Hence, it should be of great value to develop robust statistical models, based on the experimentally derived proteotypic and nonproteotypic peptides that can predict whether a given peptide is proteotypic.

The computational tools for the prediction of proteotypic peptides were developed in two recently published papers. Mallick et al. first compiled a large dataset from the yeast proteomic data produced by four different experimental platforms, then statistically extracted a few characteristic physicochemical properties that govern a peptide's proteotypic propensity, and finally developed a Gaussian mixture discriminant classifier for predicting the proteotypic propensity of a given peptide based on these properties. In contrast, Sanders et al. constructed a classifier using artificial neural networks (ANN) specific for the experimental platform, experimental protocol, and different analytical conditions of a single experiment. Thus, such models can only be applied locally.

In the present study, we develop models for predicting proteotypic peptides using support vector machines. Here, the proteotypic peptides of yeast, based on the electrospray ionization (ESI) mass spectrometer platform, were collected. The ESI mass spectrometer-based proteotypic peptides of yeast were collected from Reference [38]. For collecting the nonproteotypic peptides, we then downloaded the NIST Library of peptide ion fragmentation spectra for yeast (Version June 2006) [40], distributed by the National Institute for Systems Biology. This library contains MS/MS spectra of peptide ions also generated by ESI. The nonproteotypic peptides were finally derived by removing the proteotypic ones from the peptide sequences extracted from the NIST MS/MS library. Finally, we obtained 5,238 proteotypic and 15,077 nonproteotypic peptides. The length distribution of the two classes of peptides is shown in Figure 8.4. Then 21 length-specific datasets were compiled using the peptides ranging from 7 to 27 amino acids in length because the peptides with length <7 or >27 occupied only a very small percentage.

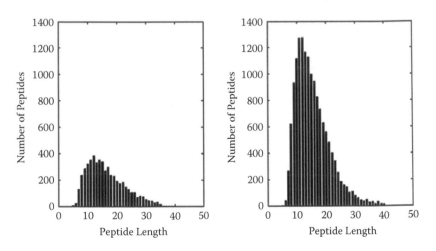

Figure 8.4 The length distribution of proteotypic peptides (left) and nonproteotypic peptides (right).

The peptide sequence has to be digitized at first in order to build recognition models with the machine-learning method. Here we employ the same method as proposed in Reference [38] to characterize the peptide sequence using a vector based on the properties of the amino acid in the Amino Acid Index 1 (AAindex1, Release 9.1) [41] which includes the physicochemical and biochemical indices of the 20 amino acids. In addition to AAindex1, the percent composition of each amino acid (20 features) is also calculated as the descriptor of the peptide. With the AAindex1 and the computed percent composition of each amino acid, a peptide sequence can then be digitized in the following procedure. (1) For each property, the numeric value of each amino acid is summed and averaged; (2) the vector consisting of the average value of each property is used to digitize the peptide sequence. For instance in Table 8.2, the peptide "EIQTAVR"can be expressed as a vector [0.998 0.427 0.966].

SVMs are a popular kernel-based method for data mining and knowledge discovery based on the structural risk minimization (SRM) rule that stems from the frame of statistical learning theory and was originally developed for the pattern recognition problem. In SVMs, the most commonly used kernel is the radial basis function (RBF).

For a given dataset, assume that each sample is denoted by x_i with the class label y_i. $x_i \in R^n$, $y_i \in (-1,1)$, $i = 1,2,3, \ldots , N$. Here, x_i is an n-dimensional vector with corresponding y_i equal to 1 if it belongs to a positive class or -1 if negative. Then the SVM algorithm can be employed to train a model with the mathematical formula:

$$y = sgn(f(x)) \tag{8.1}$$

In this study, we generalize the formula (8.1) into the form below:

$$y = sgn(f(x) - \theta), \ \theta \in [min(f(x)), max(f(x))] \tag{8.2}$$

where θ is called the decision boundary and "sgn" stands for the sign operator . It's obvious that formula (8.1) is a special case of formula (8.2) when θ is set to 0. Given a trained classifier $f(x)$, the FPR (type I error) and FNR (type II error) can be calculated and also different θ. As θ increases, FPR will decrease but FNR will increase. So this poses the problem of which θ is the best choice. In the peptide/protein identification of shotgun proteomics, the mainstream is to control the FPR at a given level, for example, 0.05. But this will inevitably cause many false negatives to be left out. From the viewpoint of statistical decision theory, a decision (here meaning the choosing of θ) should be made only when the overall loss caused by type I error and type II error achieves the minimum. That is, the researcher should pay attention to the loss caused by both types of

Table 8.2 The Digitalization Strategy of the Peptide Sequence

Accession Number[a]	Peptide sequence (EIQTAVR)							Total	Average
	E	I	Q	T	A	V	R		
ARGP820101	0.470	2.220	0.000	0.050	0.610	1.320	0.600	19.950	0.998
BHAR880101	0.497	0.462	0.493	0.444	0.357	0.386	0.529	8.536	0.427
LIFS790102	0.590	2.600	0.280	0.590	1.000	2.630	0.680	19.310	0.966

Note: ARGP820101: hydrophobicity index; BHAR880101: average flexibility indices; LIFS790102: conformational preference for parallel beta-strands.

[a] The accession number of a physicochemical property in the amino acid index database.

errors rather than purely to the reduction of FPR. Thus we propose the following loss function with the form:

$$F = \lambda\,\pi 2\,FPR + (1 - \lambda)\,\pi 1\,FNR, \lambda \in [0, 1] \tag{8.3}$$

In formula (8.3), π_1 and π_2 are the prior probability of the positive samples and negative samples, respectively. For example, if there are 20 positive and 80 negative samples in the training dataset, then $\pi_1 = 0.2$ amd $\pi_2 = 0.8$. λ is an empirically defined loss factor that characterizes the relative loss caused by FPR. For instance, if λ is set to 0.9, then $1 - \lambda$ is 0.1. This means that the loss caused by FPR is eight times bigger than that caused by FNR. By this definition, F stands for the sum of loss caused by both false positive and false negative. Once λ is fixed, we can optimize θ in formula (8.2) to minimize F. Simultaneously the FPR and FNR are also obtained when F is minimized. Finally the optimized classifier can be applied to predict whether a new peptide is proteotypic with estimated FPR. The overall strategy is shown in Figure 8.5.

For each of the 21 length-specific datasets, the peptide string is first digitized using the method proposed in Reference [38]. Then the Fisher score for each descriptor is calculated. The 10 descriptors with the highest Fisher scores are selected for classifier construction. In the present work, the most commonly used radial basis function (RBF) is chosen as kernel and the SVM classifier is built using the free software package LIBSVM [42]. Then only the regularization parameter C and the RBF kernel width σ have to be selected to construct SVM models. As is known, cross-validation is the state of the art for evaluation of the model's prediction ability and robustness. For a dataset, we take the ten fold cross-validation accuracy as the objection function, then use the genetic algorithm (GA) to optimize the two parameters C and σ in order to maximize cross-validation accuracy globally. Finally for the dataset of each peptide length (PL), an SVM classifier is trained with the maximal ten fold accuracy. The results are shown in Table 8.3, indicating that the minimum and maximum of the ten fold cross-validation accuracy are 0.724 and 0.830, respectively. And the overall average accuracy is 0.801. These results satisfy the sample complexity of the compiled datasets because the derived proteotypic peptides in our dataset come from different experimental platforms, such as PAGE-ESI, MUDPID-ESI, ICAT-ESI, and so on. P. Mallick et al. [38] have experimentally proved that proteotypic peptides from different experimental platforms have different physicochemical properties although there exist some properties in common. Thus the heterogeneity of the data decreases the distinguishability between proteotypic and nonproteotypic peptides.

Two recently published papers also concern the prediction of proteotypic peptides. Not only the datasets but also the evaluation methods for

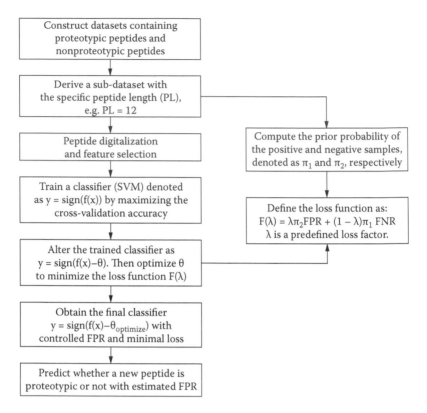

Figure 8.5 The strategy for constructing classifier and further controlling the FPR.

the trained classifiers are different from ours; it's very hard to compare the results in the present work with those reported in the literature. Mallick et al. compiled a very large training set with different lengths from multiple proteomics experiments and different experimental platforms, selected the most discriminating characteristic physicochemical properties of peptides by the Kullbach–Leibler (KL) distance and the Kolmogorov–Smirnov (KS) distance, then constructed a Gaussian mixture likelihood discriminant classifier. Their models can predict the proteotypic peptides for any protein from any organism, for a given platform, with >85% cumulative accuracy. They also report their results by area under curve (AOC) of the receiver operator characteristics (ROC) curve.

In contrast, Sanders et al. deem it necessary to construct exprimental condition-specific classifiers rather than classifiers based on datasets from different experimental instruments or analytical conditions such as Mallick's. So they described a methodology to develop artificial neural

Table 8.3 The Ten Fold Cross-Validation Accuracy and the SVM Parameters of the 21 SVM Classifier for Each of the 21 Datasets

PL	C	G	Accuracy	PL	C	G	Accuracy
7	1.48	29.67	0.754	18	30.05	116.81	0.825
8	37.31	18.87	0.797	19	19.11	50.87	0.821
9	5.90	9.00	0.793	20	30.09	177.49	0.822
10	44.71	245.14	0.801	21	3.25	10.24	0.795
11	53.65	204.08	0.830	22	5.36	21.15	0.813
12	26.51	18.76	0.823	23	234.32	1.33	0.724
13	0.17	36.05	0.816	24	27.51	153.41	0.747
14	168.22	942.51	0.824	25	3.09	40.86	0.778
15	3.00	15.55	0.820	26	142.25	63.47	0.812
16	2.03	14.65	0.822	27	42.98	248.86	0.810
17	148.66	190.92	0.794	—	—	—	—

PL = peptide length, C = regularization, G = genetic algorithm.

networks specific for the experimental platform, experimental protocol, and different analytical conditions of a single experiment. Thus, such models can only be applied locally. They also further demonstrate the need to construct dataset-specific classifiers because a classifier trained on one dataset usually has poor predictive performance when tested by another heterogeneous dataset. They report their results by ten fold cross-validation accuracy for the two datasets used in their study. In our work, the ten fold cross-validation accuracy is also used to report the SVM classifier results. The results are satisfactory considering the dataset's complexity.

To discover the most informative properties that can distinguish proteotypic peptides from nonproteotypic ones, the frequency of each property, defined as the times for a given variable to be selected in all the 21 SVM classifiers by the Fisher rule, is calculated. (See Figure 8.6.) It can be seen that some properties are selected with high frequency and others with low frequency. For instance, the highest frequency is 21 which means that the corresponding properties are selected in each of the 21 SVM classifiers. But the lowest frequency is 0 which indicates that these variables are not discriminating enough because they don't enter any SVM classifier.

The 10 most frequently selected properties together with the accession number and brief description in the AAindex database are listed in Table 8.4. It can be found that these properties are related to amino acid composition, hydrophobicity (ratio of buried and accessible molar fractions), side chain parameter, residue frequency, and so on of

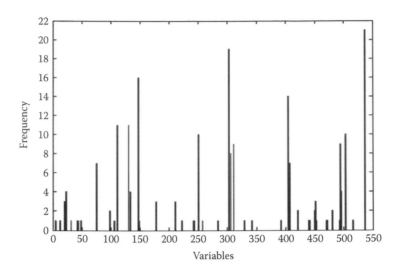

Figure 8.6 The selected frequency of each variable by Fisher rule in the 21 SVM models.

Table 8.4 The Most Frequently Selected 10 Properties in the 21 SVM Classifiers

Rank	Accession Number	Description
1	—[a]	Cysteine percent composition
2	RACS820104	Average relative fractional occurrence in EL(i) (Rackovsky–Scheraga, 1982)
3	KRIW710101	Side chain interaction parameter (Krigbaum–Rubin, 1971)
4	AURR980102	Normalized positional residue frequency at helix termini N''' (Aurora–Rose,1998)
5	JANJ790101	Ratio of buried and accessible molar fractions (Janin, 1979)
6	GRAR740101	Composition (Grantham, 1974)
7	BAEK050101	Linker index (Bae et al., 2005)
8	PRAM820103	Correlation coefficient in regression analysis (Prabhakaran–Ponnuswamy, 1982)
9	GEOR030104	Linker propensity from 3-linker dataset (George–Heringa, 2003)
10	RACS820112	Average relative fractional occurrence in ER(i-1) (Rackovsky–Scheraga, 1982)

[a] This cysteine percent composition was calculated by our group and not included in the AAindex.

peptides. These properties are related to several experimental conditions that influence the peptide's proteotypic propensity. For instance, hydrophobicity is concerned with the peptide's retention behavior in the chromatographic column of HPLC. The amino acid composition may influence the protein's digestion by trypsin. As a result, we can conclude that to a large extent these properties govern the proteotypic propensity for a peptide. By comparison we find that the 21 SVM classifiers share some properties in common but no two SVMs have the identical set of properties. This observation indicates that, for peptides of different lengths, there exist not only the general but also the distinct physicochemical properties that govern the propensity for a given peptide to be proteotypic. Thus it can be summarized that building a length-specific classifier is necessary for predicting proteotypic peptides.

In a rational and empirical manner, the proposed strategy is employed here for controlling the FPR. It aims at minimizing the λ-driven loss function $F = \lambda \pi_2 FPR + (1 - \lambda) \pi_1 FNR$, $\lambda \in [0, 1]$. The dataset with PL = 12 is applied here as an example to elucidate how FPR can be controlled. For the trained SVM classifier $f(x)$, a series of θs are uniformly sampled from the region $[\min(f(x)), \max(f(x))]$ in this case. In our case, 500 θs are sampled. Then with the given λ, F can be calculated at each θ. At last the so-called F value curve can be obtained by plotting F against θ. Two F value curves at $\lambda = 0.1, 0.5$, and 0.7 are shown in Figure 8.7. It can be seen that a minimum point (point A, B, and C) exists in each F-value curve marked by a diamond sign. The θ at the lowest point is denoted as $\theta_{optimize}$. Then the FPR can be directly obtained at $\theta_{optimize}$. Therefore an optimized SVM classifier can be obtained as y = sgn($f(x) - \theta_{optimize}$) with estimated FPR. In this case the FPR at $\theta_{optimize}$ for $\lambda = 0.1, 0.5$, and 0.7 are 0.6814, 0.1049, and 0.0157, respectively. This result indicates that FPR will decrease as λ increases. It should be pointed out here that λ is an empirical parameter determined by the relative loss brought about by type I and type II error. For example in the peptide identification of the bottom-up proteomics, researchers usually filter out a large number of false negatives by setting a high threshold, such as Xcorr in SEQUEST. Thus λ should be set to a high value in order to control the FPR at a low level.

In general, the SVM classifiers learned from the unbalanced datasets have the potential to distinguish the proteotypic peptides from the nonproteotypic ones with a certain accuracy. From the point of view of experimental proteomics, this may help the biologist or proteomist to discover new proteins in a mixture or identify the experimentally unobserved proteins that are referenced to participate in a canonical biochemical pathway.

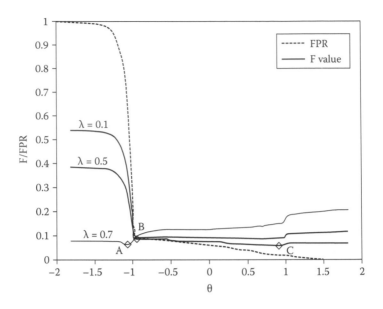

Figure 8.7 The F-value curve (solid line) and the FPR (broken line) curve at different θ. Three F-value curves are shown with λ = 0.1, 0.5, and 0.7. There exists a minimum point (points A, B, and C) in each F-value curve marked by a diamond sign. Then the FPR can be directly obtained at the θ corresponding to the lowest point of the F-value curve.

8.5 Biomarker discovery in metabolomics using support vector machines

One of the goals of metabolomics studies is to discover informative metabolites or biomarkers that can be used for disease diagnosis, drug discovery, and so on. To date, statisticians, chemometricians, and biologists have developed a large number of methods [21,43–46] that can be utilized to reveal the informative metabolites, the changes of which may reflect potential metabolic disorders in living systems [47–50]. In this section, we show the application of SVM to biomarker discovery in a type 2 diabetes mellitus (T2DM) study. These data contain the plasma samples of 45 patients and 45 controls. After background correction [51], two-way data resolution, and integration [52,53], we have finally identified 21 metabolites. Details of the data can be found in our previous work [17,18].

To identify an optimal subset of discriminating metabolites, we used the aforementioned RFE and MIA to conduct variable selection. We also used linear kernelized SVM. The penalizing factor C was chosen by ten fold cross-validation. For MIA, the two tuning parameters Q and N were

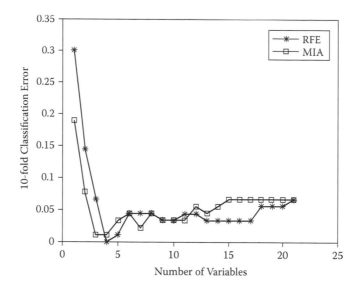

Figure 8.8 The ten fold cross-validated classification error against the number of selected variables by RFE and MIA on the type 2 diabetes mellitus data.

set to 15 and 2,000, respectively. By running RFE and MIA on these data, the ten fold cross-validated prediction errors against the number of metabolites were computed and shown in Figure 8.8. The subset identified by RFE contained four variables, achieving the classification error 0.00. For MIA, it finally selected an optimal subset of three metabolites. The ten fold cross-validated error is 0.011, slightly larger than that of RFE. There are two metabolites identified by both methods. They are α-linolenic acid (C18:3n-3) and eicosapentaenoic acid (EPA, C20:5n-3). These two metabolites are shown to have biological meaning (details not shown here) in our previous work [17,18].

8.6 Some remarks

With the development of life sciences and technology, there is growing complexity of the data produced by a wide variety of instruments in OMICS studies. The analysis of the huge amount of data plays a more and more important role in modern scientific research. In this context, many statistical or machine-learning methods are introduced into life science, aiming at extracting useful information and gaining a better understanding of biological systems. Among these learning methods, SVM has proven promising. It's expected that SVM will find more and more successful applications in the challenging OMICS research.

References

1. http://en.wikipedia.org/wiki/Genomics.
2. Alexey, I. N., Olga, V., and Ruedi, A. 2007. Analysis and validation of proteomic data generated by tandem mass spectrometry. *Nat. Meth.*, 4(10):787–797.
3. Sebastian, K. and Anne-Claude, G. 2003. Towards quantitative analysis of proteome dynamics. *Nat. Genet. Supple.*, 33:311–323.
4. Thomas, K. and Giulio, S.-F. 2007. Mass spectrometry-based functional proteomics: From molecular machines to protein networks. *Nat. Meth.*, 4(10):807–815.
5. Sebastian, K., and Anne-Claude, G. 2007. Towards quantitative analysis of proteomics dynamics. *Nat. Biotechnol.*, 25(3):298–300.
6. Katherine, H.-W. and Dorothee, K. 2007. Dynamic personalities of proteins. *Nature*, 450(13):964–972.
7. Aebersold, R. and Mann, M. 2003. Mass spectrometry-based proteomics. *Nature*, 422(6928):198–207.
8. Daviss, B. 2005. Growing pains for metabolomics. *Scientist*. April, 19(8):25–28.
9. Jordan, K., Nordenstam, J., Lauwers, G., Rothenberger, D., Alavi, K.M.G., and Cheng, L. 2009. Metabolomic characterization of human rectal adenocarcinoma with intact tissue magnetic resonance spectroscopy. *Diseases Colon Rectum*, 52(3):520–525.
10. Gavaghan, C.L., Holmes, E., Lenz, E., Wilson, I.D., and Nicholson, J.K. 2000. An NMR-based metabonomic approach to investigate the biochemical consequences of genetic strain differences: Application to the C57BL10J and Alpk:ApfCD mouse. *FEBS Lett.*, 484(3):169–174.
11. Alakorpela, M., Korhonen, A., Keisala, J., Horkko, S., Korpi, P., Ingman, L.P., Jokisaari, J., Savolainen, M.J., and Kesaniemi, Y.A. 1994. H-1 NMR-based absolute quantitation of human lipoproteins and their lipid contents directly from plasma. *J. Lipid Res.*, 35(12):2292–2304.
12. Makinen, V.P., Soininen, P., Forsblom, C., Parkkonen, M., Ingman, P., Kaski, K., Groop, P.H., Ala-Korpela, M., Finn, D. 2008. Study G: H-1 NMR metabonomics approach to the disease continuum of diabetic complications and premature death. *Mol. Syst. Biol.*, 4:167
13. Mierisova, S. and Ala-Korpela, M. 2001. MR spectroscopy quantitation: A review of frequency domain methods. *Nmr. Biomed.*, 14(4):247–259.
14. Tukiainen, T., Tynkkynen, T., Makinen, V.P., Jylanki, P., Kangas, A., Hokkanen, J., Vehtari, A., Grohn, O., Hallikainen, M., Soininen, H. et al. 2008. A multi-metabolite analysis of serum by H-1 NMR spectroscopy: Early systemic signs of Alzheimer's disease. *Biochem. Biophys. Res. Commun.*, 375(3):356–361.
15. Zeng, M.M., Liang, Y.Z., Li, H.D., Wang, M., Wang, B., Chen, X., Zhou, N., Cao, D.S., and Wu, J. 2010. Plasma metabolic fingerprinting of childhood obesity by GC/MS in conjunction with multivariate statistical analysis. *J. Pharmaceut. Biomed. Anal.*, 52(2):265–272.
16. Zeng, M.-M., Xiao, Y., Liang, Y.-Z., Wang, B., Chen, X., Cao, D.-S., Li, H.-D., Wang, M., and Zhou, Z.-G. 2010. Metabolic alterations of impaired fasting glucose by GC/MS based plasma metabolic profiling combined with chemometrics. *Metabolomics*, 6(2): 303–311.

17. Tan, B.-B., Liang, Y.-Z., Yi, L.-Z., Li, H.-D., Zhou, Z.-G., Ji, X.-Y., and Deng, J.-H. 2009. Identification of free fatty acids profiling of type 2 diabetes mellitus and exploring possible biomarkers by GC–MS coupled with chemometrics. *Metabolomics*, 6(2): 219–228.

18. Li, H.-D., Zeng, M.-M., Tan, B.-B., Liang, Y.-Z., Xu, Q.-S., and Cao, D.-S. 2010. Recipe for revealing informative metabolites based on model population analysis. *Metabolomics*, 6(3):353–361.

19. Yi, L.-Z., He, J., Liang, Y.-Z., Yuan, D.-L., and Chau, F.-T. 2006. Plasma fatty acid metabolic profiling and biomarkers of type 2 diabetes mellitus based on GC/MS and PLS-LDA. *FEBS Lett.*, 580(30):6837–6845.

20. Ben-Dor, A., Bruhn, L., Friedman, N., Nachman, I., Schummer, M., and Yakhini, Z. 2000. Tissue classification with gene profiles. *J. Comput. Biol.*, 7:559–584.

21. Golub, T.R., Slonim, D.K., Tamayo, P., Huard, C., Gaasenbeek, M., Mesirov, J.P., Coller, H., Loh, M.L., Downing, J.R., Caligiuri, M.A. et al. 1999. Molecular classification of cancer: Class discovery and class prediction by gene expression monitoring. *Science*, 286(5439):531–537.

22. West, M., Blanchette, C., Dressmna, H., Huang, E., Ishida, S., Spang, R., Zuzan, H., Olson, J., Marks, J., and Nevins, J. 2001. Predicting the clinical status of human breast cancer by using gene expression profiles. *Proc. Natl. Acad. Sci. USA*, 98:11462–11467.

23. Zhang, L., Zhou, W., Velculescu, V.E., Kern, S.E., Hruban, R.H., Hamilton, S.R., Vogelstein, B., and Kinzler, K.W. 1997. Gene expression profiles in normal and cancer cells. *Science*, 276(5316):1268–1272.

24. Alon, U., Barkai, N., Notterman, D.A., Gish, K., Ybarra, S., Mack, D., and Levine, A.J. 1999. Broad patterns of gene expression revealed by clustering analysis of tumor and normal colon tissues probed by oligonucleotide arrays. *Proc. Natl. Acad. Sci. USA*, 96(12):6745–6750.

25. Guyon, I., Weston, J., Barnhill, S., and Vapnik, V. 2002. Gene selection for cancer classification using support vector machines. *Mach. Learn.*, 46(1):389–422.

26. Liu, X., Krishnan, A., and Mondry, A. 2005. An entropy-based gene selection method for cancer classification using microarray data. *Bmc Bioinf.*, 6(1):76.

27. Chen, D., Cai, W., and Shao, X. 2007. Representative subset selection in modified iterative predictor weighting (mIPW)—PLS models for parsimonious multivariate calibration. *Chemometr. Intell, Lab,*, 87(2):312–318.

28. Li, H.-D., Liang, Y.-Z., Xu, Q.-S., and Cao, D.-S. 2010. Model population analysis for variable selection. *J. Chemometr.*, 24(7–8):418–423.

29. Li, H.-D., Liang, Y.-Z., Xu, Q.-S., Cao, D.-S., Tan, B.-B., Deng, B.-C., and Lin, C.-C. 2010. Recipe for uncovering predictive genes using support vector machines based on model population analysis (accepted). *IEEE/ACM Trans. Comput. Biol. Bioinf.*,

30. Mann, H.B. and Whitney, D.R. 1947. On a test of whether one of two random variables is stochastically larger than the other. *Ann. Math. Statist.*, 18:50–60.

31. Furey, T.S., Cristianini, N., Duffy, N., Bednarski, D.W., Schummer, M., and Haussler, D. 2000. Support vector machine classification and validation of cancer tissue samples using microarray expression data. *Bioinformatics*, 16(10):906–914.

32. Nguyen, D. and Rocke, D.M. 2002. Tumor classification by partial least squares using microarray gene expression data. *Bioinformatics*, 18:39–50.
33. Dettling, M. and Buhlmann, P. 2003. Boosting for tumor classification with gene expression data. *Bioinformatics*, 19(9):1061–1069.
34. Gualdrón, O., Brezmes, J., Llobet, E., Amari, A., Vilanova, X., Bouchikhi, B., and Correig, X. 2007. Variable selection for support vector machine based multisensor systems. *Sensor. Actuat. B-Chem.*, 122(1):259–268.
35. Domon, B. and Aebersold, R. 2006. Review - Mass spectrometry and protein analysis. *Science*, 312(5771):212–217.
36. Frewen, B.E., Merrihew, G.E., Wu, C.C., Noble, W.S., and MacCoss, M.J. 2006. Analysis of peptide MS/MS spectra from large-scale proteomics experiments using spectrum libraries. *Anal. Chem.*, 78(16):5678–5684.
37. Jimmy, K.E., Ashley, L.M., John, R.Y.I. 1994. An approach to correlate tandem mass spectral data of peptides with amino acid sequences in a protein database. *J. Amer. Sot. Mass. Spectrom.*, 5:976–989.
38. Mallick, P., Schirle, M., Chen, S.S., Flory, M.R., Lee, H., Martin, D., Raught, B., Schmitt, R., Werner, T., Kuster, B. et al. 2007. Computational prediction of proteotypic peptides for quantitative proteomics. *Nat. Biotechnol.*, 25(1):125–131.
39. Kuster, B., Schirle, M., Mallick, P., and Aebersold, R. 2005. Scoring proteomes with proteotypic peptide probes. *Nat. Rev. Mol. Cell Biol.*, 6(7):577–583.
40. http://www.peptideatlas.org/speclib/.
41. Shuichi, K., and Minoru, K. 2000. AAindex: Amino acid index database. *Nucl. Acids Res.*, 28(1):374.
42. http://www.csie.ntu.edu.tw/~cjlin/libsvm.
43. Tibshirani, R. 1996. Regression shrinkage and selection via the lasso. *J. Roy. Statist. Soc. B*, 58:267–288.
44. Zou, H. and Hastie, T. 2005. Regularization and variable selection via the elastic net. *J. Roy. Statist. Soc. B*, 67:301–320.
45. Wongravee, K., Lloyd, G., Hall, J., Holmboe, M., Schaefer, M., Reed, R., Trevejo, J., and Brereton, R. 2009. Monte-Carlo methods for determining optimal number of significant variables. Application to mouse urinary profiles. *Metabolomics*, 5(4):387–406.
46. Madsen, R., Lundstedt, T., and Trygg, J. 2010. Chemometrics in metabolomics: A review in human disease diagnosis. *Anal. Chim. Acta*, 659(1–2):23–33.
47. Boudonck, K.J., Mitchell, M.W., Wulff, J., and Ryals, J.A. 2009. Characterization of the biochemical variability of bovine milk using metabolomics. *Metabolomics*, 5(4):375–386.
48. Graham, S.F., Amigues, E., Migaud, M., and Browne, R.A. 2009. Application of NMR based metabolomics for mapping metabolite variation in European wheat. *Metabolomics*, 5(3):302–306.
49. Crews, B., Wikoff, W.R., Patti, G.J., Woo, H.K., Kalisiak, E., Heideker, J., and Siuzdak, G. 2009. Variability analysis of human plasma and cerebral spinal fluid reveals statistical significance of changes in mass spectrometry-based metabolomics data. *Anal. Chem.*, 81(20):8538–8544.
50. Bertram, H.C., Eggers, N., and Eller, N. 2009. Potential of human saliva for nuclear magnetic resonance-based metabolomics and for health-related biomarker identification. *Anal. Chem.*, 81(21):9188–9193.

51. Zhang, Z.M., Chen, S., and Liang, Y.Z. Baseline correction using adaptive iteratively reweighted penalized least squares. *Analyst*, 135(5):1138–1146.
52. Kvalheim, O.M. and Liang, Y.-Z. 1992. Heuristic evolving latent projections: Resolving two-way multicomponent data. 1. Selectivity, latent-projective graph, datascope, local rank, and unique resolution. *Anal. Chem.*, 64(8):936–946.
53. Liang, Y.Z., Kvalheim, O.M., Keller, H.R., Massart, D.L., Kiechle, P., and Erni, F, 1992. Heuristic evolving latent projections: Resolving two-way multicomponent data. 2. Detection and resolution of minor constituents. *Anal. Chem.*, 64(8):946–953.

Index

Printed and bound by CPI Group (UK) Ltd, Croydon, CR0 4YY

21/10/2024

01777085-0006